The UK Low Carbon Transition Plan

National strategy for climate and energy

Presented to Parliament pursuant to Sections 12 and 14 of the Climate Change Act 2008

Amended 20th July 2009 from the version laid before Parliament on 15th July 2009.

15 July 2009

£34.55

Amendments to the version laid before Parliament on 15 July 2009

The following corrections have been made:

p16-17: 7.5% energy from renewables corrected to sit in 2018, not 2016; over 10% corrected to sit in 2018; and extra circle added to read 15% in 2020.

p49: chart 8 title "fossil fuels" corrected to "gas".

p52, p55, p57: 2% corrected to 22%.

p78: chart 1 title "%15" corrected to read "13%".

p99: first paragraph, fifth line "bills" corrected to read "prices".

p158: "the Government and" has been inserted between "...EU rules on state aid" and "the Carbon Trust will work to make..."

p200: savings from SAFED training for bus drivers corrected to read 0.1 in 2010, not -0.1.

p207: zero carbon homes corrected to read 0.2 (2018), 0.3 (2019), 0.3 (2020) and 0.3 (2021).

pp208-9: the figures in the Total row have been corrected.

p212: Table A8, contribution from workplaces and jobs corrected to read 64 in 2013; contributions from farms, land and waste corrected to read 70 (2008 & 2009), 71 (2010 & 2011), 73 (2015, 2016 & 2017), 69 (2018 & 2019) and 70 (2020, 2021 & 2022).

p214: Chart A3 title and labels corrected, consistent with descriptive text.

A few other minor typographic errors have also been corrected.

ISBN: 9780108508394

Foreword by Secretary of State

The transition to a low-carbon economy will be one of the defining issues of the 21st century. This plan sets out a route-map for the UK's transition from here to 2020.

With the discipline of carbon budgets, legally binding limits on emissions, we plan to drive change in every area: the way we generate energy, the way we heat our homes and workplaces and the way we travel.

In Britain, as our own reserves in the North Sea decline, we have a choice: replace them with ever-increasing imports, be subject to price fluctuations and disturbances in the world market and stick with high carbon; or make the necessary transition to low carbon, right for climate change, energy security and jobs.

The transition gives us the chance to lead the clean industries of the future. In demonstrating the technology to capture carbon dioxide and lock it away, for example, we can lay the pipes and the infrastructure for new, sustainable industrial hubs, and we gain the engineering knowledge to win contracts installing it in other countries.

Across business, we can build up the skills to be more resource-efficient. Like the internet, saving carbon can become part of how business is done: every financial officer knowing their savings and liabilities from carbon, every builder having the skills to build in a way which saves energy.

There will be costs to the transition. But they are far outweighed by the costs if we didn't act and faced the expense of adapting and coping with dangerous climate change. The task for government is to minimise the costs of the low carbon path and spread them fairly. That's why we are committed to drive forward energy efficiency, ensure tough regulation and provide extra support for the most vulnerable.

The transition will show that Britain, while not the biggest country in the world or the largest polluter, can lead in preventing the worst effects of climate change.

The new predictions from the Met Office and other scientists, the most detailed yet, show that the impacts of climate change are not just an issue for other countries and future generations, but an urgent issue for Britain.

Making the transition will take strategic action by government and a comprehensive plan. This is that plan. It shows sector-by-sector what savings can be achieved and how every department across government will take responsibility.

But the changes cannot be done by government alone. There has been good progress – Britain has already cut 21% of emissions since 1990 but the move to carbon budgets signals a change of pace, and the scale of the task is enormous.

So alongside the country's low carbon transition plan, every business, every community will need to be involved. Together we can create a more secure, more prosperous low carbon Britain and a world which is sustainable for future generations.

Ed Miliband

Secretary of State of Energy and Climate Change

Contents

Executive summary

Summary

This White Paper sets out the UK's first ever comprehensive low carbon transition plan to 2020. This plan will deliver **emission cuts of 18% on 2008 levels by 2020** (and over a one third reduction on 1990 levels).

Key steps include:

For the first time, **all major UK Government departments have been allocated their own carbon budget** and must produce their own plan.

Getting 40% of our electricity from low carbon sources by 2020 with policies to:

- **Produce around 30% of our electricity from renewables** by 2020 by substantially increasing the requirement for electricity suppliers to sell renewable electricity.
- **Fund up to four demonstrations of capturing and storing emissions from coal power stations.**
- Facilitate the building of new nuclear power stations.

Clarifying that **Ofgem**, in its job to protect consumers, both current and future, **should help tackle climate change and ensure security of supply**.

Making homes greener by:

- **Channelling about £3.2 billion to help households become more energy efficient** by increasing the current programme by 20% between 2008 and 2011 and then extending it to the end of 2012.
- Rolling out smart meters in every home by the end of 2020.
- **Piloting "pay as you save" ways to help people make their whole house greener** – the savings made on energy bills will be used to repay the upfront costs.
- **Introducing clean energy cash-back schemes** so that people and businesses will be paid if they use low carbon sources to generate heat or electricity.
- **Opening a competition for 15 towns, cities and villages** to be at the forefront of pioneering green innovation.

Helping the most vulnerable by:

- Creating **mandated social price support** at the earliest opportunity with increased resources compared to the current voluntary system. The Government is minded to focus new resources particularly on older pensioners on the lowest incomes.
- **Piloting a community-based approach to delivering green homes in low income areas**, helping around 90,000 homes.
- **Increasing the level of Warm Front grants** so most eligible applicants can receive their energy saving measures without having to contribute payment themselves.

Helping make the UK a centre of green industry by supporting the development and use of clean technologies, including up to **£120 million investment in offshore wind and an additional £60 million to cement the UK's position as a global leader in marine energy**.

Transforming transport by cutting average carbon dioxide emissions from new cars across the EU by 40% on 2007 levels, supporting the largest demonstration project in the world for **new electric cars**, and sourcing 10% of UK transport energy from sustainable renewable sources by 2020.

The first ever formal framework for tackling emissions from farming.

Producing a longer term roadmap for the transition to a low carbon UK for the period 2020 to 2050 by next spring **and a vision for a smart grid**.

Setting out the Government's assessment of the outlook for energy security.

This White Paper sets out the UK's transition plan for becoming a low carbon country: cutting emissions, maintaining secure energy supplies, maximising economic opportunities, and protecting the most vulnerable.

The challenge

If the world continues emitting greenhouse gases like carbon dioxide at today's levels then average global temperatures could rise by up to 6°C by the end of this century. This is enough to make extreme weather events like floods and drought more frequent and increase global instability, conflict, public health-related deaths and migration of people to levels beyond any of our recent experience. Heat waves, droughts, and floods would affect the UK too.

To avoid the most dangerous impacts of climate change, average global temperatures must rise no more than 2°C, and that means global emissions must start falling before 2020 and then fall to at least 50% below 1990 levels by 2050.

The UK is calling for an ambitious global agreement at UN talks in Copenhagen in December 2009. The Government's approach to this deal is set out in The Road to Copenhagen published in June 2009.

To encourage action, the EU, which represents the UK in these UN talks, has promised to cut its emissions to 20% below 1990 levels by 2020, and by 30% if other countries play their part. The UK will make an above average contribution to meeting these, reflecting our relatively high income.

The EU has also created the world's largest emissions trading scheme, which could form the basis of a global system to cut emissions and help fund emissions cuts in developing countries. And it is exemplifying the kinds of further targeted action that is needed by supporting renewable energy, testing new technologies and setting standards to cut emissions from cars and other products.

Changes in our climate mean that two out of every three people on Earth could experience water shortages by 2025

Driving the transition

We all need to play our part in making these changes. If we get it right, we will have a better quality of life, improved long-term economic health, new business opportunities in a fast-growing global sector, and, by reducing our reliance on fossil fuels, greater security of future energy supplies.

But the transition is not without its challenges. We will need to drive major changes to the way we use and supply our energy and in doing so it is critical that our supplies continue to be safe, secure and reliable. We need new investment in low carbon infrastructure and to manage the risks associated with our increasing dependence on energy imports at a time when competition for global energy supplies is intensifying. Over time, energy costs will rise, so the Government will be vigilant in ensuring affordable prices and helping the most vulnerable.

The UK has made good progress so far. Emissions have already fallen 21% below 1990 levels, nearly double what was promised at Kyoto, and over 800,000 people are employed in low carbon businesses. But there is more to do.

Dynamic, competitive markets, a strategic role for Government, and active communities will be needed to bring about the transition to low carbon.

To drive this transition, the Government has put in place the world's first ever legally binding target to cut emissions 80% by 2050 and a set of five-year "carbon budgets" to 2022 to keep the UK on track.

This White Paper for the first time sets out how these budgets will be met – so that by 2020 UK emissions will be 18% below 2008 levels and over one third below 1990 levels. This will mean emissions falling faster than before: emissions have fallen about 1% a year since 1990, and will now fall 1.4% a year. The UK will go even further if other countries sign up to an ambitious global agreement.

Chart 1
The plan will reduce emissions in every sector

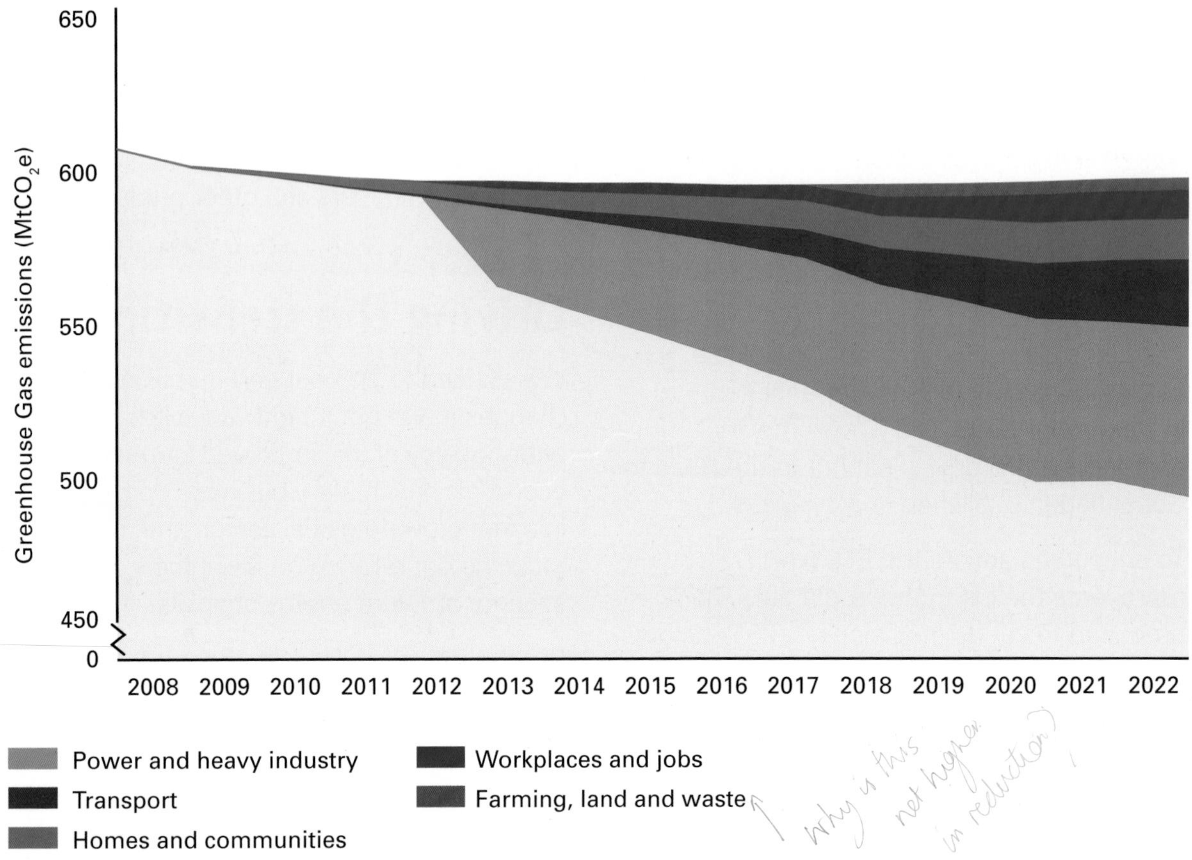

Source: Department of Energy and Climate Change

Note: The impact of policies prior to the 2007 Energy White Paper is included in the baseline; without these policies, UK emissions would be higher.

For the first time, UK Government departments have been allocated their own carbon budgets. Every delivery department will produce its own plan to show how it will stay within its carbon budget. If the Government fails to ensure that the UK lives within its carbon budgets then it will have to purchase credits from abroad.

One of the key ways the UK will achieve its carbon budgets is through a commitment in law to get 15% of all our energy – for electricity, heat and transport – from renewable sources by 2020. **This White Paper sets out the Government's plan to achieve this seven-fold increase**, with more detailed plans in the *Renewable Energy Strategy* published in parallel with this Transition Plan.

This White Paper sets out the Transition Plan to 2020 for transforming our power sector, our homes and workplaces, our transport, our farming and the way we manage our land and waste, to meet these carbon budgets, secure energy supplies, maximise economic opportunities and protect the most vulnerable. These are detailed in the sections below.

To deliver these goals the Government will:

- Drive decarbonisation, by providing a carbon price, supporting the new technologies and infrastructure we need and helping households and businesses overcome barriers to low carbon choices.
- Secure energy supplies by ensuring a supportive climate for the substantial new investment needed to bring forward low carbon infrastructure, and maximise the economic production of oil and gas from the North Sea to help secure the continued fossil fuel supplies needed during the transition.

The plan will keep our energy supplies safe and secure

- Help UK low carbon and energy businesses to grow.
- Protect consumers, in particular the most vulnerable.
- Help businesses manage the costs of tackling climate change and help everyone adapt to climate impacts.
- Protect the environment by making the most of measures which bring wider environmental benefits and minimising impacts where they are unavoidable.

Chart 2

The main policies driving emission reductions are the EU Emissions Trading System, energy efficiency policies, and increased use of renewable energy for heat and transport

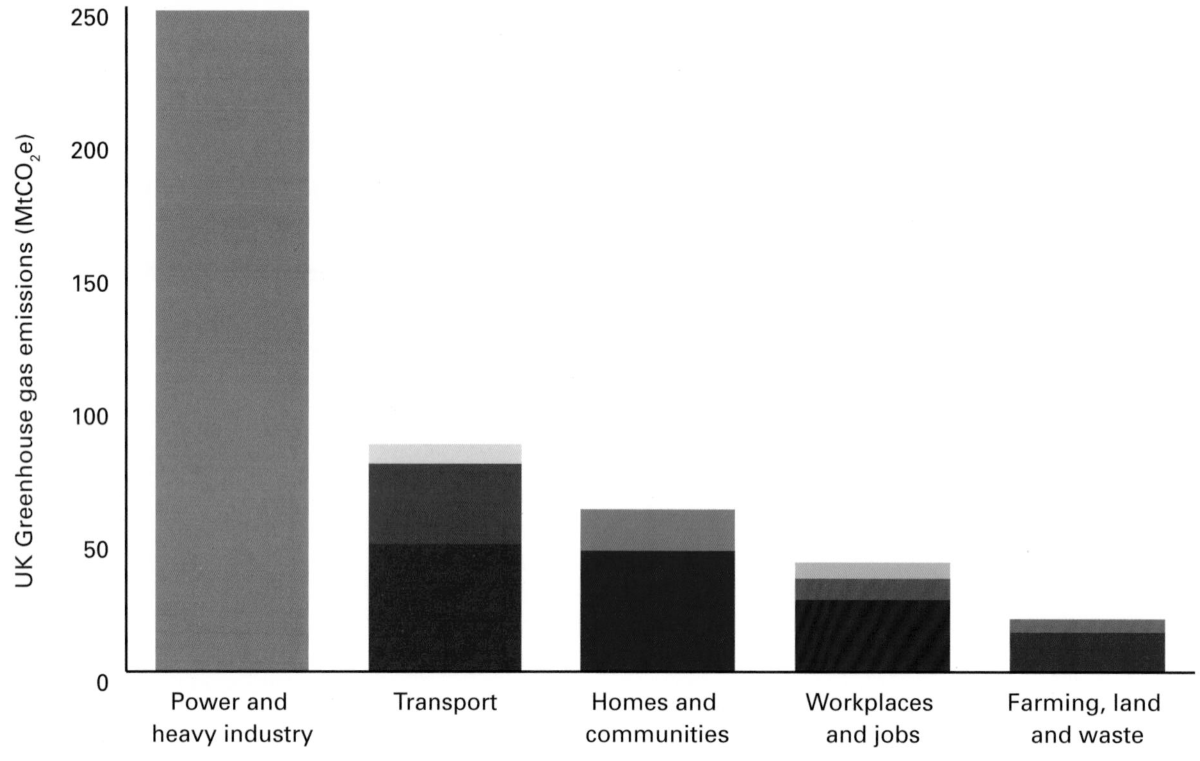

- European Union Emissions Trading System
- New vehicle CO_2 policies
- Additional renewable transport fuels
- Low carbon buses, car improvement technologies, driver training, illustrative rail electrification of 750km of track
- Energy efficiency, smart metering, Community Energy Saving Programme , and zero carbon homes
- Clean energy cashback (renewable heat incentive)
- Clean energy cashback (renewable heat incentive)
- Climate Change Agreements and other policies
- Carbon Reduction Commitment and other policies
- Farming (crop management, manure management etc.)
- Waste policies (diverting waste from landfill, increased landfill tax)

Source: Department of Energy and Climate Change

Note: The impact of policies prior to the 2007 Energy White Paper is included in the baseline; without these policies, UK emissions would be higher.

Transforming our power sector

Three quarters of our electricity comes from coal and gas, and the power and heavy industry sector accounts for 35% of UK emissions. But by 2050 virtually all electricity will need to come from renewable sources, nuclear or fossil fuels where emissions are captured and safely stored for the long term. Electricity is likely to be used more extensively for heat and transport, so we will probably need more than today.

The plan to 2020 will secure power supplies and cut emissions from power and heavy industry together by 22% on 2008 levels, over half of the savings needed to meet carbon budgets, so that by 2020 around 40% of our electricity will come from low carbon sources.

A key tool for delivering this is the EU Emissions Trading System which places a limit on emissions from electricity and heavy industry (and, from 2012, aviation too).

But this alone will not be enough to enable the rapid development and use of low carbon technologies. So the Government is providing additional tailored support:

- **Renewables: renewable electricity will increase to around 30% by 2020**, a five-fold increase.The Government is now **launching the Office for Renewable Energy Deployment** to help make this change happen, including developing supply chains to bring jobs to the UK, and is publishing a short list of possible Severn Tidal projects.
- **Nuclear:** the Government is streamlining the planning and regulatory approvals processes for new nuclear power stations. It is currently assessing sites where developers would like to bring new nuclear power stations into operation by 2025,

The UK will need a bigger, smarter electricity grid for the future and this assessment will be included in a draft National Policy Statement for nuclear power, which the Government will consult on later in 2009.

- **Carbon capture and storage:** the Government has announced plans for a mechanism to **support up to four demonstrations** and a proposed requirement for any new coal power stations to demonstrate this technology, and measures to prepare the UK for roll out if the technology is proven. A consultation on these plans was launched in June 2009.

We will also need a **bigger, smarter electricity grid**. The Government has therefore endorsed industry plans to increase grid capacity, taken action to speed up connection of renewable electricity to the grid and is supporting development of new technologies which could enable the grid to work better in the future. **The Government will later this year publish a high level vision for a future smart grid** and subsequently a plan for delivering this.

The scale and pace of the transformation that the Government expects to see in the electricity sector means that it needs to be alert to new challenges. Ensuring security of supply is a particular challenge because of the lead times for building new power stations and the requirement for significant capital investment. The Government is therefore creating a supportive climate for timely investment in a diverse mix of low carbon technologies. The Government is also ensuring that the market and regulatory framework can adapt to cope with the different characteristics of low carbon electricity generation technologies.

The Government believes that the risk to security of electricity supplies over the next decade is low but that the scale and pace of change required will test the electricity market during the transition to a low carbon economy. The Government will shortly issue a call for evidence on electricity markets to further explore these issues.

Transforming our homes and communities

Three quarters of the energy we use in our homes is for heating our rooms and water, most of which comes from gas-fired boilers. Together this accounts for 13% of the UK's greenhouse gas emissions, and by 2050 emissions from homes need to be almost zero by using energy more efficiently and using more low carbon energy.

The plan to 2020 will cut emissions from homes by 29% on 2008 levels, introduce further measures to protect the most vulnerable, and improve the security of our gas supplies, a third of which is used in our homes.

Two thirds of the homes we will live in by 2050 have already been built, so we will need to make our existing homes much more energy efficient and heat and power them from low carbon sources. So the Government is:

Energy efficiency improvements could save households an average of £300 a year

- **Increasing the obligation on energy suppliers to help households reduce emissions and save energy, the Carbon Emissions Reduction Target, by 20%** between April 2008 and March 2011, so that about £3.2bn will be invested. Six million households have already been helped since 2002. But now the **obligation will be extended to the end of 2012**, which is expected to benefit 1.5 million additional households.
- **Introducing a community-based approach to delivering significant energy efficiency treatments to 90,000 homes in low-income areas**, the Community Energy Saving Programme.
- **Piloting a move from upfront payment to 'pay as you save' models of long-term financing for energy saving**, so it will be more affordable to make the changes needed to make the whole house low carbon.
- **Introducing "clean energy cash-back" schemes** so that people, businesses and communities will be paid if they use low carbon sources to generate heat or electricity. A household with well-sited solar panels could receive over £800, plus bill savings of around £140 a year.

We all need to act, and the Government is helping encourage collective action by:

- **Rolling out smart meters** in every home by the end of 2020, which will enable people to understand their energy use, maximise opportunities for energy saving, and offer better services from energy companies.
- Encouraging the provision of smart displays for existing meters now, benefitting some 2-3 million households, and launching **a new personal carbon challenge with rewards and incentives for saving energy**.
- **Developing more proactive services from the Energy Saving Trust** to provide households with information and advice when it is most likely to be useful.
- Launching a competition for 15 communities to be at the forefront of pioneering green initiatives.

Bigger changes will require new approaches, so the Government is:

- Consulting on **requiring Energy Performance Certificate** ratings for rented properties to be put on property advertisements; and consulting on extending access to the performance information to help target efficiency offers and support.
- Requiring new homes to be built to higher environmental standards and from 2016 all new homes will be 'zero carbon homes'.
- Considering how best to deliver significant 'whole house' energy saving treatments in the longer term, setting out the strategy this autumn.

The policies in this Transition Plan will increase household energy bills. By 2020, the additional impact of all the policies in this plan, relative to today, is equivalent to approximately a 6% increase from current energy bills. When previously announced climate policies are included this figure is 8%.

The Government intends to clarify Ofgem's remit. The Government continues to believe that effective competition remains the central way by which consumers' interests can be protected. However, there are contexts in which means other than competition may be a preferable way to protect their interests. **The Government proposes to amend the legislation to make this clearer, building on the existing legislation**.

Tackling fuel poverty is a priority for the Government and it has set itself a target to end fuel poverty, as far as reasonably practicable, in vulnerable households by 2010 and in all other households by 2016.

To help the most vulnerable, Government has already put in place a £20 billion package of support with payments for older and more vulnerable people, and subsidised energy efficiency measures and new heating systems. The Warm Front programme fits or repairs a central heating system every minute of every working day in vulnerable households across England. And more than 800,000 vulnerable households in, or at risk of, fuel poverty currently receive discounts and other help with their energy bills as part of a voluntary agreement negotiated between Government and the energy companies.

Now, in addition, the Government is:

- Creating **mandated social price support** at the earliest opportunity with increased resources compared to the current voluntary system. The Government is minded to focus new resources particularly on older pensioners on the lowest incomes.
- **Increasing the level of Warm Front grants** so most people receiving benefits get their energy saving measures without having to contribute payment themselves.
- Working to ensure that fuel poor households can benefit from new low carbon schemes, such as the Renewable Heat Incentive, to help reduce bills.

Most homes rely on gas for heating. This Transition Plan will reduce UK gas demand across the economy by 27% compared to 2008 levels. But the UK remains heavily dependent on gas and so the Government is helping to ensure that the UK has reliable supplies. The UK is expected to rely on net imports to meet around 45% of its net gas demand in 2020, compared to the level of around 60% expected without the Government's policies. But the diversity of its gas supplies has helped the UK to remain largely unaffected by international disputes. Future security of supply will also require that the UK improves its capacity to import and store gas and develops strategic partnerships with international gas suppliers.

The Government will shortly issue a commentary on the outlook for the security of UK gas supplies. Malcolm Wicks MP has reviewed how the UK can maintain secure energy supplies during the transition to a low carbon economy, and his report will be published in the coming months.

Transforming our workplaces and jobs

The changes we need to make to 2020 and beyond will transform our workplaces and our whole economy. Our workplaces are responsible for 20% of UK emissions. By 2050 all of our workplaces will need to be using less energy and making use of clean energy to reduce greenhouse gas emissions and potentially save billions of pounds each year.

The plan to 2020 will cut emissions from our workplaces by 13% on 2008 levels, and build the UK's position as a global centre of green manufacturing in low carbon sectors such as offshore wind, marine energy, low carbon construction and ultra-low carbon vehicles.

The Government will help reduce emissions from workplaces by:

- Including high carbon industries in the EU Emissions Trading System, which will save around 500 million tonnes of carbon dioxide a year across the EU by 2020.

- Providing financial support and incentives for business and the public sector to save energy and invest in low carbon technologies including the Climate Change Levy and Climate Change Agreements, Carbon Reduction Commitment and low cost loans and grants for businesses and the public sector.
- Providing advice to help all workplaces change through the Carbon Trust, Business Link, the Waste and Resources Action Programme and Envirowise.

But there will be costs from this transition: the additional impact in 2020 of the policies in this Transition Plan, relative to today, is equivalent to approximately a 15% increase in current energy bills for businesses consuming a medium amount of energy. When previously announced climate policies are included this is 17%. The Government is working to ensure that competitive energy markets deliver low cost energy and EU frameworks are fair to business.

But there are also huge opportunities for UK businesses to take part in the £3 trillion world low carbon market that will employ over 1 million people in the UK by the middle of the next decade.

To help make the UK a world centre of the green economy, the Government is:

- **Investing in research and development of new low carbon technologies, including by using the £405 million announced in April 2009 to deliver a major boost to technologies where the UK has the greatest potential**, as described in more detail in the *UK Low Carbon Industrial Strategy* published in parallel with this Transition Plan.
  - This includes **up to £120 million of investment in offshore wind, and investment of up to an additional £60 million to cement the UK's position as a global leader in marine energy and help develop the South West of England as the UK's first Low Carbon Economic Area**.
  - **The plan will also deliver support for a smart electrical grid, ultra-low carbon vehicle infrastructure and exploration of deep geothermal power**.
- Helping businesses to take up new opportunities by strengthening delivery of support for research and development, and taking action to help employees develop new skills.
- Supporting businesses through the global financial crisis and facilitating access to up to £4 billion of new capital for renewable and other energy projects from the European Investment Bank.

There could be 1.2 million people in the UK working in green sectors by 2015

Transforming transport

A fifth of our greenhouse gas emissions come from transport but by 2050 they will have to be radically reduced by using energy more efficiently and moving to more low carbon forms of energy.

The plan to 2020 will cut emissions from transport by 14% on 2008 levels and secure the oil supplies needed during the transition to a low carbon country.

As set out in *Low Carbon Transport: a Greener Future* published in parallel with this Transition Plan, the first step is to improve the fuel efficiency of new conventional vehicles, so the Government is:

- **Cutting average carbon dioxide emissions from new cars across the EU to 95g/km by 2020**, a 40% reduction from 2007 levels.
- Ensuring that the Government leads by example by setting targets for government departments and their agencies to procure new cars for administrative purposes that meet the EU standard for 2015 in 2011, four years early
- **Pressing the EU to require new vans** to be more efficient.
- **Investing up to £30 million over the next two years to deliver several hundred low carbon buses.**

We must move away from petrol and diesel in the long term. So the Government is testing out options for the radically different technologies needed by:

- **Demonstrating 340 new electric and lower carbon cars** on the UK's roads, the largest project of its kind in the world.
- **Providing help worth about £2,000 to £5,000 per vehicle towards reducing the price of ultra-low carbon cars, from 2011**, and up to £30 million to support the installation of electric vehicle charging infrastructure in six or so cities across the UK.

The Government is improving cycle storage facilities in railway stations

- Committing to source 10% of UK transport energy from sustainable renewable sources by 2020.

Cutting transport emissions is not just about changing technologies. We all need to make low carbon travel choices, and the Government is helping by:

- **Launching a competition** for the country's first Sustainable Travel City, building on projects in towns which saw reported car trips fall by 9%, walking increase by 14% and cycling increase by 12%.
- Investing £140 million in promoting cycling in England in 2008-11, and a **new £5 million investment in improving cycle storage at rail stations**.

Emissions from international flights and ships are growing. The Government is pushing hard for an international agreement to cut them, the only truly effective way to do so, and meanwhile is:

- Putting a cap on emissions from all flights arriving at or leaving from European airports by including them in the EU Emissions Trading System from 2012.
- Introducing a target to limit UK aviation emissions to below 2005 levels by 2050, despite forecast growth in passenger demand, which is likely to be met through more efficient engines and other new technologies, and supported by government policies such as changes to airport passenger duty.

In the longer term the UK needs to reduce its dependence on oil for transport but it will still be an important fuel for some time to come and the Government needs to help ensure that the UK has safe and secure supplies of the oil products it requires. The Government's approach is to maximise the economic exploitation of the UK's own oil reserves, to work with other countries to ensure a well-functioning global oil market, and to improve UK fuel infrastructure.

For the first time farming and land use emissions will be included in a framework for tackling emissions

Transforming farming and managing our land and waste sustainably

Farms, changes in land use and waste contribute 11% of UK greenhouse gas emissions. We need to find ways of emitting less while safeguarding our environment and producing food sustainably. The equivalent of around 37 billion tonnes of carbon dioxide is currently locked into natural reservoirs of carbon like soils and forests – we need to carefully manage our land to keep these stores locked away.

The UK now recycles or composts a third of its waste, but we need to do more because rubbish dumped on landfills continues emitting greenhouse gas for many decades.

The plan to 2020 will cut emissions from farming and waste by 6% on 2008 levels through:

- **Encouraging English farmers to take action themselves to reduce emissions to at least 6% lower than currently predicted by 2020**, through more efficient use of fertiliser, and better management of livestock and manure.
- Support for anaerobic digestion, a technology that turns waste and manure into renewable energy.
- Reducing the amount of waste sent to landfills, and better capture of landfill emissions.

The plan will also **encourage private funding for woodland creation.**

Changes over the next

Power

2009

- Wind (onshore and offshore) produces over 4GW of power
- Government publishes a high level vision for a future smart grid
- Third round of leases for 25GW offshore wind sites awarded
- Shortlist of possible Severn Tidal schemes published
- Pay as you save pilots start

2010

- New planning regime under Infrastructure Planning Commission begins
- Anticipated first deployment of wave and tidal energy demonstration projects under the Marine Renewables Deployment Fund
- Reforms to the Renewables Obligation are introduced
- Government makes a decision on Severn Tidal scheme
- Government introduces new long-term grid access rules

2011

- Levy on electricity suppliers to fund CCS demonstration projects in place
- Commissioning of Wave Hub energy testing centre in Cornwall and first deployment of wave energy devices
- Expansion of wave and tidal energy testing sites in Northumbria and Orkney completed

2012

2013

- The cap for the EU Emissions Trading System starts to be tightened every year from now
- The power sector starts paying for every tonne of carbon emitted by purchasing allowances in EU Emissions Trading System auctions
- Construction of first new nuclear power stations expected to be underway

2014

- First UK commercial scale carbon capture and storage demonstration intended to be operational
- Larger-scale wave and tidal energy generation (>10MW) starts to be deployed

2012: 4%of total energy (including power, heat and transport) to come from renewable sources

2014: Over 5% of total energy from renewable sources

Homes and communities

2009

- Community Energy Saving Programme starts trialling "whole house" treatments in low income areas

2010

- About 95% of social housing stock in England meets improved Decent Homes standard
- Clean energy cash back for electricity starts (Feed in Tariffs)
- Building Regulations improve energy efficiency by 25% compared to 2006 regulations

2011

- Energy wasting traditional light bulbs are no longer sold
- 6 million homes will have been insulated under the Carbon Emissions Reduction Target, Decent Homes, the Community Energy Saving Programme and Warm Front
- Clean energy cash back for renewable heat starts in April (the Renewable Heat Incentive)

2012

- The Community Energy Savings Programme will have helped 90,000 homes to improve their energy efficiency in 100 areas around Great Britain

2013

- Building Regulations improve energy efficiency by 44% compared to 2006 regulations

2014

10 years

Share of 2018-22 emissions savings

- Power and heavy industry 54%
- Homes and communities 13%
- Workplaces and jobs 9%
- Transport 19%
- Farming, land and waste 4%

2015

The EU will have selected 12 carbon capture and storage demonstration projects for support across the EU

2016

2017

2018

Plans show first new nuclear power station operational

2019

2020

Around 30% of electricity is generated from renewable sources

Up to four carbon capture and storage demonstration projects operational in the UK

7.5% of total energy to come from renewable sources

Over 10% of total energy to come from renewable sources

15% of total energy to come from renewable sources

2015

All lofts and cavity walls in Great Britain insulated where practical

400,000 homes will benefit from "whole house" packages of energy efficiency and low carbon energy per annum

2016

All new homes zero carbon

2017

2018

2019

2020

By end of 2020 every home in Great Britain will have a smart meter

1.8 million homes will benefit from "whole house" packages of energy efficiency and low carbon per annum

Around 12% of heat is generated from renewable sources, equivalent to supplying 4 million households based on current heating demand

Changes over the next

Workplaces and jobs

2009	2010	2011	2012	2013	2014
880,000 people work in the green sector Government provides £1.4 billion of targeted support for low carbon industries in the world Central Government departments take on carbon budgets for their own estate and operations	Carbon Reduction Commitment introductory phase begins Central government buildings will be 15% more efficient than in 1999/00	First sale of allowances for the Carbon Reduction Commitment for 2010 and 2011 in April New period for Climate Change Agreements begins		Emissions from large businesses and public sector become capped under the Carbon Reduction Commitment Current Climate Change Agreements end	

Transport

2009	2010	2011	2012	2013	2014
Almost 340 ultra-low emission cars on the road in the coming 18 months, the largest project of its kind in the world Delivery of several hundred low carbon buses over 2009 and 2010 Improved cycle storage facilities at up to 10 major railway stations during 2009-10	Renewable Transport Fuel Obligation is amended or replaced to deliver renewable transport goals Following High Speed Two's report to Government at the end of 2009, Government intends to consult on proposals for a new high speed rail line between London and the West Midlands	The Government provides £2,000-5,000 per vehicle to help reduce the cost of ultra low carbon cars Government departments and agencies meet target to procure new cars for administrative purposes that meet EU standard for 2015 by2011	All flights arriving in or departing from European airports part of the EU Emissions Trading System 500,000 more children trained to ride safely through the Bikeability programme Government to set an environmental target for train operators for the period 2014-19	Fuel suppliers are required to ensure that 5% of road transport fuel comes from renewable sources by 2013/14	New Super Express trains which are greener and less noisy are rolled out from 2014
2009	2010	2011	2012	2013	2014

Farming, land and waste

2009	2010	2011	2012	2013	2014
Anaerobic digestion implementation plan published, Government responds later in the year Government publishes consultation on landfill bans	Government publishes: options for action if agricultural emissions do not reduce fast enough; improvements to food labelling; and plans for tighter control of emissions from landfill		Government reviews voluntry action by farmers to address agricultural emissions and decides whether to intervene		

10 years

Share of 2018-22 emissions savings

- Power and heavy industry 54%
- Homes and communities 13%
- Workplaces and jobs 9%
- Transport 19%
- Farming, land and waste 4%

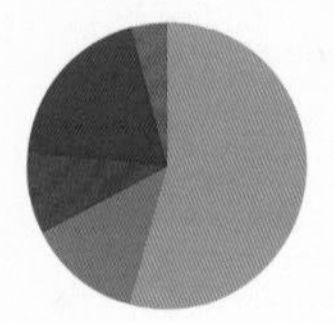

2015

1.2m people could be working in the green sector

Low carbon economy could be worth £150m a year in UK and £4.3bn a year globally

The NHS expects to have reduced its carbon footprint by 10% compared to today

2016

All new schools proposed to be zero carbon (subject to consultation and confirmation)

2017

Climate Change Agreements extension to 2017 ends

Carbon Reduction Commitment first capped Phase ends. Cap for second Phase set

2018

New nuclear power stations could create or sustain up to 9000 jobs during the course of construction and operation (including supply chains)

Carbon Reduction Commitment second capped phase starts

Government ambition for all new public sector (non-domestic) buildings to be zero carbon (subject to further work)

2019

Government ambition for all new non-domestic buildings to be zero carbon from this date (subject to consultation and confirmation)

2020

Up to half a million additional jobs in the UK renewable energy sector, including supply chains

Central government departments (and wider public sector) will have cut their greenhouse gas emissions by 30% from 1999/00

2015

Average level of emissions from new cars sold in Europe is 130g CO_2/km

2016

2017

2018

2019

2020

Average carbon dioxide emissions from new cars in Europe will be 95g CO_2/km – representing a 40% improvement from 2007 levels

10% of transport energy to come from sustainable renewable sources

2015

2016

2017

2018

Agriculture is efficient, competitive, and climate-friendly

Very little biomass is landfilled, emissions are tightly controlled, and material formerly landfilled is used for renewable energy, compost and fertilizer

2019

2020

Chapter 1

The challenge

Summary

Action on climate change is urgently needed to prevent widespread human suffering, ecological catastrophes, and political and economic instability. In 2006, the *Stern Review* concluded that the costs of uncontrolled climate change could be in the range of 5% to 20% of global gross domestic product (GDP) per year, averaged over time.[1]

Action to reduce greenhouse gas emissions can offer improved energy and international security, a better environment, new economic opportunities and a fairer society. If emissions are reduced in a low cost way to a level that avoids the most dangerous risks of climate change, then the costs of acting could be as low as 1% to 2% of global GDP by 2050.[2]

Achieving this and preventing climate change will require a strong global framework; the UK is arguing for an international deal at Copenhagen in December that is ambitious, effective and fair.

Like every country, the UK must take action now. We need to plan our own low carbon transition. This will mean major changes to the way we use and supply energy. We need to ensure secure supplies of energy throughout the transition and to deliver change in a way that maximises economic opportunities and protects the most vulnerable.

Extreme weather conditions will become more frequent as the climate changes

1. Direct costs: not including costs such as those related to political and economic disruption, and impacts of migrations
2. Based on a target stabilisation concentration of greenhouse gases of 450-550 parts per million

The urgency of action

The consensus of scientists spanning over 130 countries, is now overwhelming: human activities are causing global climate change.[3] The burning of fossil fuels, changes in land use, and various industrial processes are adding heat-trapping gases, particularly carbon dioxide (CO_2), to the atmosphere. There is now roughly 40% more CO_2 in the atmosphere than there was before the industrial revolution. Such high levels have not been experienced on earth for at least 800,000 years and in all likelihood not for the last three million years.

The effects of these additional greenhouse gases can already be seen today. Global average temperatures have risen by 0.75°C since about 1900 (see chart 1), with consequences for both the environment and people's lives.

Figure 1
Increasing concentrations of greenhouse gases are changing the climate

1. Sunlight passes through the atmosphere and warms the earth.

2. Infrared Radiation (IR) is given off by the earth. Most IR escapes to outer space and cools the earth.

3. But some IR is trapped by gases in the air and this reduces the cooling effect.

Source: Met Office (2009)

3. Intergovernmental Panel on Climate Change *Fourth Assessment Report* (2007)

Chart 1
Global temperature rises are linked to the growth in emissions of CO2 and other greenhouse gases

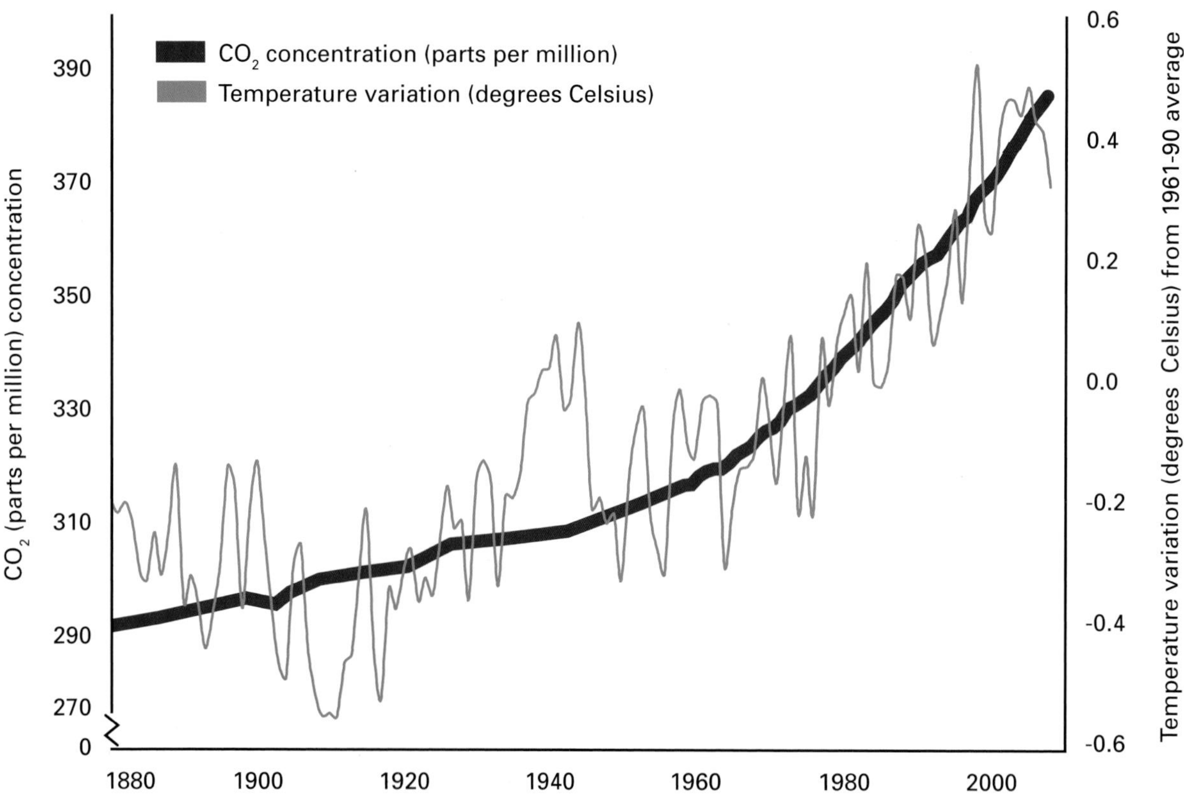

Source: CO_2 data pre-1958 from ice cores and post-1958 from the Mauna Loa observatory. Temperature data from Met Office.

- Summer Arctic sea ice continues to decline, with the smallest amounts ever recorded occurring in 2007 and 2008 (see figure 2).
- Global sea levels have already risen by 10cm over the last 50 years, and are conservatively projected to increase by between 18 and 59 cm by the end of this century, increasing the flood risk for some major coastal cities.
- Ocean acidity is rising as a result of greenhouse gases, and is already having a detrimental impact on the many ocean animals that build shells of calcium carbonate, including many tropical reef-building corals, molluscs and crustaceans such as lobsters.
- Plants and animals from warmer climates now inhabit previously cooler regions. In some marine and freshwater systems, changes in the abundance of algae, plankton and fish are associated with rising water temperatures.
- Rising temperatures are already affecting aspects of human health, such as heat-related deaths during heat waves and changes in the spread of infectious disease.
- Within the UK, the nine hottest days on record for central England have all occurred in the last 15 years with 2006 being the warmest year recorded.[4]

4. UK Climate Projections (2009)

Figure 2
Summer Arctic sea ice continues to decline

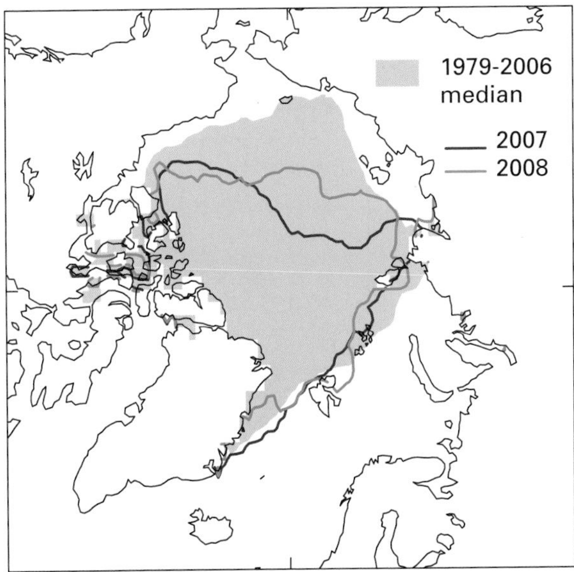

Source: Met Office (2008)

If climate change continues unchecked, the additional consequences for the UK will be severe. The *UK Climate Projections*, published in 2009, gives the most detailed predictions yet of the most likely major changes to the UK's climate in the absence of action to cut global emissions. For example, by 2080 we can expect:

- More droughts: In Yorkshire and Humber, summer rainwater could drop by a quarter.
- More flooding: In the South East, winter rains and snow could increase by almost a third.
- Damaging heat waves: In 2003, an increase in average temperature of just two degrees led to 35,000 extra deaths across Europe. Such summers could soon count as being cooler than average. In the West Midlands, without a reduction in emissions, summer averages could increase by almost five degrees.

Across the world, the consequences of failing to control emissions would be worse still (see figure 4). 70% of Africans rely on agriculture for a living; by 2020, climate change is predicted to cut some rain-fed farming harvests by half in the continent.[5] Two of every three people on earth could experience water shortages by 2025.[6] Three-

Climate change will make extreme weather events like these more common in the UK

In the 2003 heat wave, 35,000 people died prematurely across Europe

In the summer 2007 downpours, 55,000 properties were flooded and 350,000 people were left without mains water

5. Intergovernmental Panel on Climate Change Fourth Assessment Report, Working Group 2 (2007)
6. UN World Water Development Report 3 (2009)

Figure 3
The UK could experience summers that are 3°C hotter by 2080

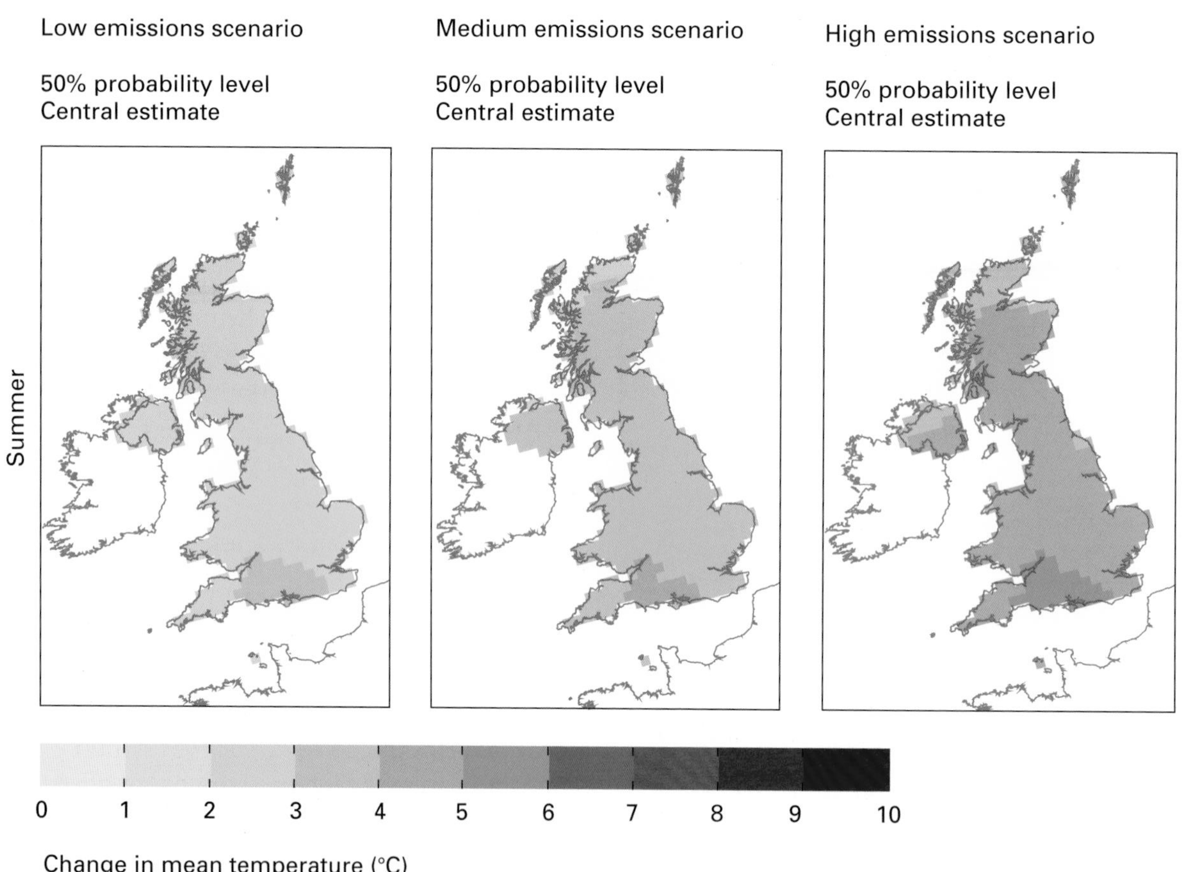

Temperature increase relative to 1961 baseline, projections based on different 'business as usual' scenarios. The high scenario is based on heavier reliance on fossil fuels, whereas the low scenario assumes a strong move away from fossil fuels.

The Intergovernmental Panel on Climate Change definition of the central estimate is "as likely as not". It is important to consider the entire range of projections to gain a full picture of the distribution of uncertainty, and to encompass all possible changes in climate, as indicated by the current science.[7]

Source: © *UK Climate Projections* (2009)

quarters of a billion people in Asia rely on water that melts steadily from glaciers in the Himalayas; but the glaciers are already melting faster than in any other part of the world, and if the present rate continues, they are likely to disappear by 2035.[8] By 2050, 200 million people could be rendered homeless by rising sea levels, floods and drought.[9] By 2080, an extra 600 million people worldwide could be affected by malnutrition and an extra 400 million people could be exposed to malaria.[10]

The world's poorest countries will bear the brunt of a changing climate. A rise in temperature of a few degrees would slow or even reverse their development, and directly prevent poor people from lifting themselves from poverty.

7. For further information, see http://ukcp09.defra.gov.uk
8. Intergovernmental Panel on Climate Change *Fourth Assessment Report*, Working Group 2 (2007)
9. Myers, N *Environmental Refugees: An emergent security issue* 13th Economic Forum, Prague (2005)
10. UNDP Human Development Report *Fighting Climate Change: Human Solidarity in a Divided World* (2007-8)

Figure 4

If the world continues producing high levels of greenhouse gases, there will be significant impacts globally and in the UK

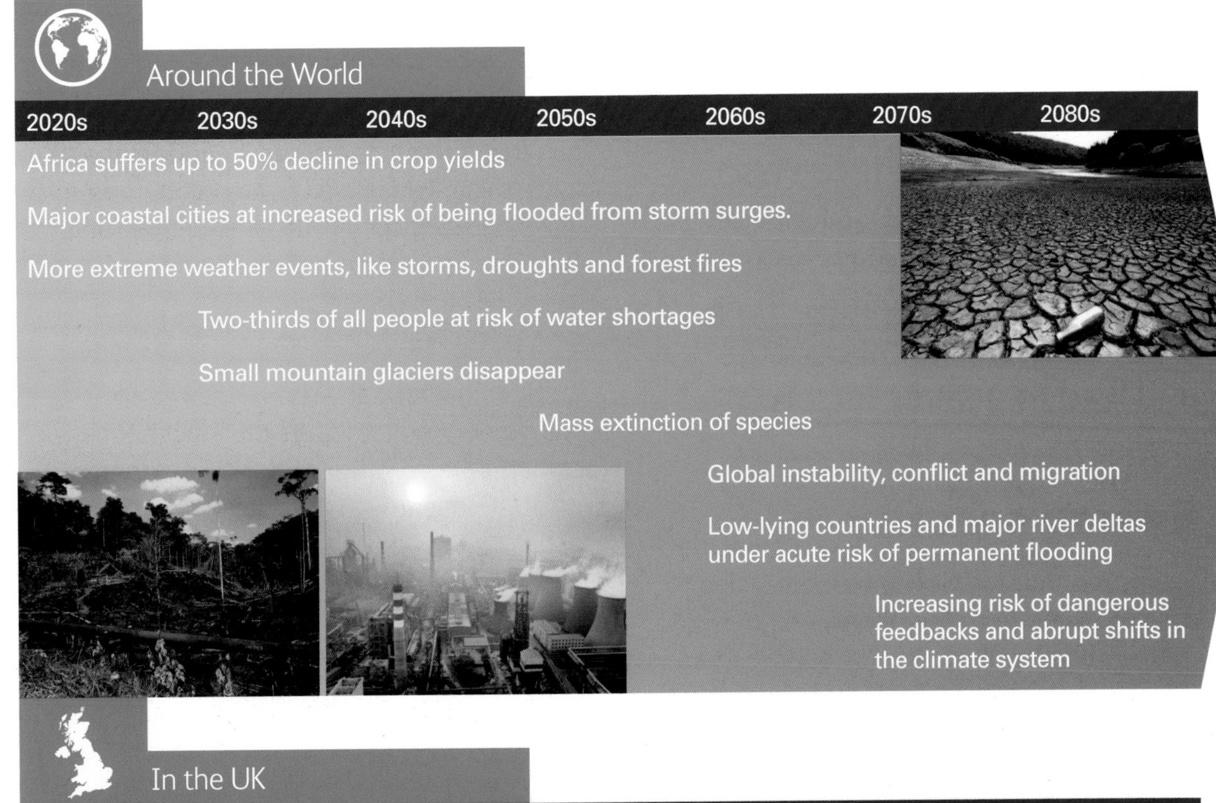

In the UK

2020s	2030s	2040s	2050s	2060s	2070s	2080s

Severe risks to national infrastructure: coastal defences, sewage system, rail

Modest increase in agriculture yields

Floods like those of 2007 will be frequent

Several UK species struggle to adapt.

Heat wave of 2003 will be 'normal' by the 2050s

By 2080, 4°C rise in average summer temperatures

Temperature of the hottest summer days up by possibly 10°C

Up to 40% reduction in summer rainfall

Source: Intergovernmental Panel on Climate Change Fourth Assessment Report (2007), UK Climate Projections (2009), Stott et al Human contribution to the European heat wave of 2003, Nature (2004)

A stable climate and a better life

These impacts are not inevitable. Early action can prevent the worst excesses of climate change. It can also improve the security of Britain's energy supplies, create new economic opportunities, bring wider environmental benefits, and lead to a fairer society.

Energy security

The action that the Government is taking to cut emissions from the energy sector is good for the security of our energy supplies too. Global energy demand is forecast to increase by around 45% between 2006 and 2030, with almost 80% of this increase coming from fossil fuels.[11]

Without action, the UK would rely even more on imported fossil fuels and would have greater exposure to global energy price fluctuations (see chart 2), especially when demand recovers as the world emerges from the global economic downturn.

In 2008 the UK imported about 25% of the gas that it used. Projections suggest that by 2020 this could rise to around 60%[12]. But with the measures in this plan, especially those which help to decarbonise our electricity supplies and increase our heat efficiency, we can reduce this to 45%.[13]

Substantial private sector investment will be needed to deliver this new low carbon infrastructure. The UK Government wants to ensure reliable, secure supplies of energy during the transition, including from the fossil fuels we will continue to rely on in the future. The pathway to delivering low carbon energy supplies is a necessary one but will need careful design to manage the energy security risks.

Preventing climate change and securing energy supplies go hand-in-hand

11. International Energy Agency World Energy Outlook (2008): this is according to the IEA's reference scenario, taking account of all policy measures introduced by Governments at the time of publication
12. Without Government policies to reduce dem and
13. Reflecting implementation of measures contained in this plan; see technical annex

Chart 2
International oil prices are highly volatile

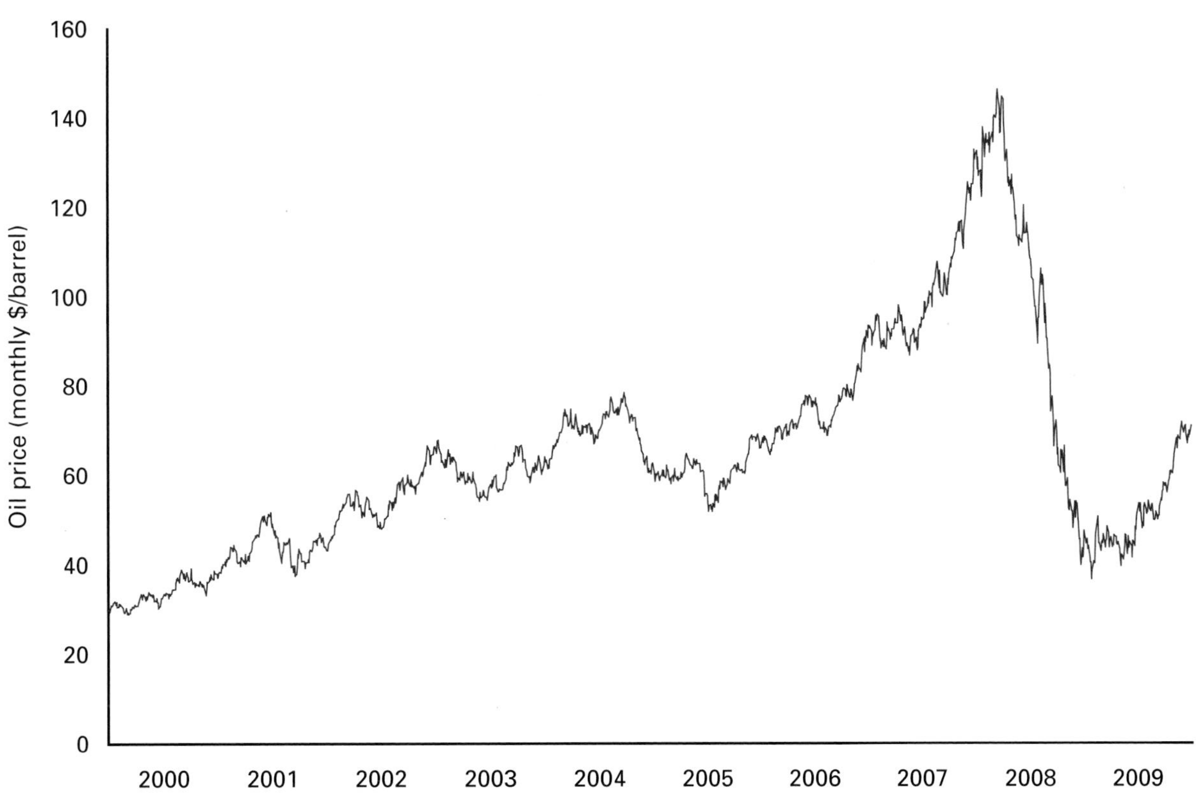

Source: Intercontinental Exchange www.theice.com/homepage.jhtml (1 month-Brent)

Economic opportunities

The economic case for urgent action is compelling. First, the costs of not acting are much greater than the costs of acting: in 2006, Lord Stern, former Chief Economist of the World Bank, estimated that if the world did nothing to tackle climate change then changes to global weather systems could inflict costs equivalent to between 5% and 20% of global GDP a year, averaged over time.[14] He has since said that new science research published since the *Stern Review* suggested the review underestimated the risks and dangers of inaction.[15] On the other hand, if the world takes action to reduce emissions in a low cost way to a level consistent with avoiding dangerous climate change, Stern estimates the cost as 1% to 2% of global GDP by 2050.[16]

Second, acting on climate change will stimulate innovation and new technologies to help businesses reduce energy costs, and will provide employment in 'green industries'. Already, low carbon and environmental goods and services are worth an estimated £3 trillion worldwide.[17] This fast growing global market will generate huge demand for the technologies, goods and services that will allow countries across the world to make the transition to a low carbon economy.

There will be costs to business in adapting to this world, and every part of the economy will need to change the way it operates, but for those who seize them, the opportunities are

14. Stern Review on the Economics of Climate Change (2006)
15. Speech made in Copenhagen at conference Climate Change: Global Risks, Challenges and Decisions (March 2009)
16. Based on a target stabilisation concentration of greenhouse gases of 450-550 parts per million
17. Innovas Low Carbon and Environmental Goods and Services , (2009)

Low carbon industries offer exciting global business opportunities

equally vast. The UK, which already has a low carbon and environmental sector worth £106 billion a year, has the chance to become a world centre of green manufacturing, and the Government is committed to help business to grasp that opportunity.

There will be broader benefits for international and national security.[18] The effects of climate change are one of the key drivers of national security threats, exacerbating weaknesses and tensions around the world.[19] The changes that scientists are predicting in our climate unless we can stabilise it will worsen poverty, have a significant impact on global migration patterns, and risk tipping fragile countries into instability, conflict and state failure.

Fairness

Tackling climate change provides many opportunities to create a fairer society. Some of these actions will require upfront investment, but they will bring long-term benefits, in particular for the poor and most vulnerable. Better insulation can make homes more comfortable in winter, and more green spaces can provide a cooler environment in summer (reducing the need for air conditioning) and also renew run-down areas. Innovative transport options and more flexible working patterns can make it easier to use public transport or to cycle, and reduce the need for commuting. Health improvements will also be gained by the reduction in air pollution, for example from car engines. Grasping the opportunities for solving problems together can build a sense of community, wherever people live.

Preventing dangerous climate change is self-evidently in Britain's national interest: in the interests of citizens who want a healthier life, a more prosperous economy and greater international and energy security. But more fundamentally, given the impact and permanency of the effects, it is a moral imperative. Future generations will pay the price if we fail to rise to this challenge.

There will be many lifestyle advantages of a low carbon economy

18. Shared Responsibilities: A National Security Strategy for the UK, The Final Report of the IPPR Commission on National Security in the 21st Century (2009)
19. The National Security Strategy of the United Kingdom: Update 2009 - Security for the Next Generation (2009)

Global climate change needs a global response

For all these reasons, the UK is arguing for an international agreement that matches the scale of the challenge.

Under the United Nations, countries from around the world will meet in Copenhagen in December 2009. The UK argues that the ambition of the deal must match the science. If temperature rises exceed 2°C compared to pre-industrial times, the risks of major, irreversible changes and feedback effects increase. Even if the world could stabilise levels of greenhouse gases tomorrow, warming of at least 1.4°C by 2100 is almost certain due to the time lag between emissions and temperature rise.[20]

In an effort to keep the increase below 2°C, the UK is calling for developed countries as a group to reduce their emissions by at least 80% by 2050 compared to 1990 levels, including stretching mid-term targets on a pathway to getting there. The Intergovernmental Panel on Climate Change's analysis suggests developed countries should collectively reduce their emissions by 25-40% below 1990 levels by 2020. But even if developed countries could reduce emissions to zero, the world as a whole would still not avoid temperature increases above 2°C.

Chart 3
To avert the most dangerous impacts of climate change, global emissions must peak within the next decade

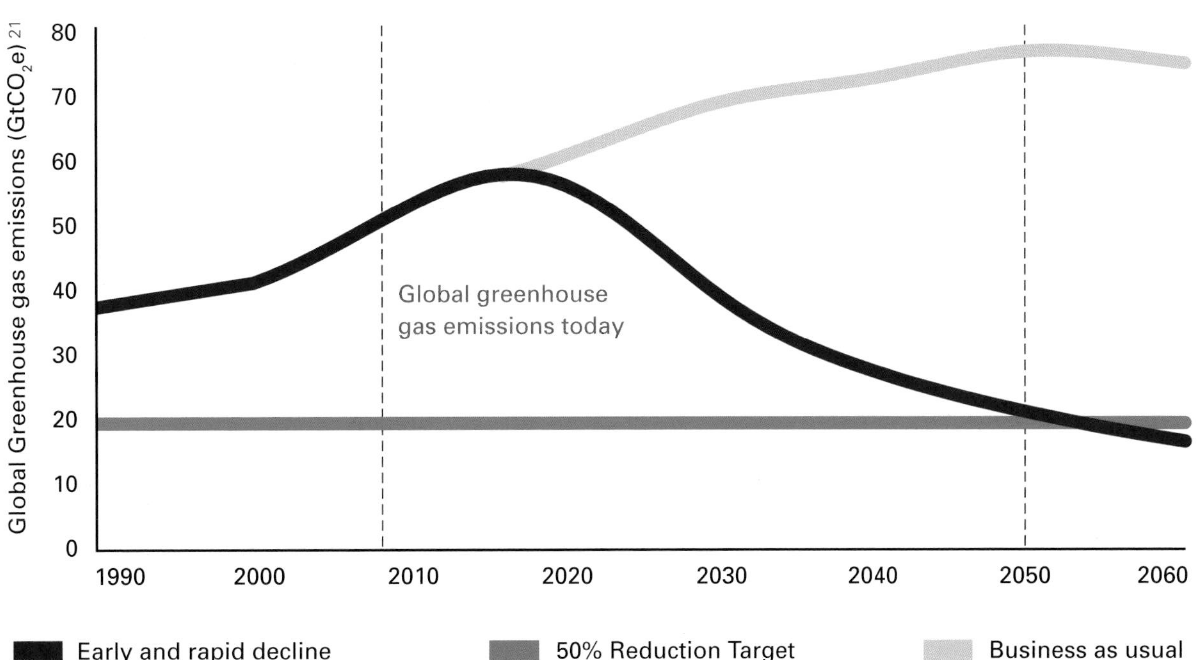

Source: An unmitigated emissions scenario from the Intergovernmental Panel on Climate Change *Special Report on Emissions Scenarios* (2000)

20. Relative to pre-industrial temperatures, or 0.65°C relative to today.
21. CO2e is Carbon Dioxide Equivalence, which describes for a given greenhouse gas the quantity of carbon dioxide that would have the same global warming potential when measured over a specified timescale

Figure 5

Limiting temperature rises to 2°C means stabilising the concentration of greenhouse gases, which means a big cut in global emissions

For...	we need to...	this means...
a 50:50 chance of staying below 2°C relative to pre-industrial temperatures	stabilise concentrations at 450ppm CO_2e	cutting global emissions by at least 50% of 1990 levels by 2050

The greenhouse gas emissions of many developing countries are due to rise rapidly in the short term

That is because of the predicted growth of developing countries. The UK is therefore calling for action from developing countries that would see global emissions peak by 2020 and fall to less than 50% of 1990 levels by 2050.

An ambitious agreement must also ensure that all major sources of emissions are included, including those from international aviation and shipping and deforestation.

A global agreement must also be effective: if countries are to meet ambitious emission reduction commitments then investment must flow to where it has the most effect. The UK therefore supports the development of a global carbon market, including the expansion and linking of emissions trading systems as part of a broader economic transformation needed for a low carbon world.

Finally, the UK argues the global deal must be fair. A priority for Copenhagen will be to put in place finance to support developing countries to cut their emissions and adapt to climate change.

The 2009 publication, *The Road to Copenhagen: The UK Government's case for an ambitious international agreement on climate change* sets out in more detail the UK's priorities and how it aims to achieve them.

Act on Copenhagen

www.actoncopenhagen.gov.uk is the official UK Government website presenting information on the climate change negotiations. It will act as the domestic and international hub for information and communications in the lead up to the UN talks in Copenhagen in December 2009.

The Copenhagen conference will be a key stage in building global agreement to act on climate change

Figure 6
A global deal to tackle climate change should be ambitious, effective and fair

Ambitious	Effective	Fair
• Strong commitments to reduce emissions from both developed and developing countries • Cover all sources of emissions, including aviation and shipping and deforestation	• Establish a reformed and expanded global carbon market • Enable low carbon and adaptive technologies to be developed and disseminated • Provide robust monitoring, reporting and verification arrangements	• Support developing countries to adapt to the climate change that is now unavoidable • Provide sufficient finance, technological assistance and capacity building to enable developing countries to take action on both mitigation and adaptation • Establish governance structures that strengthen the voice of developing countries

Source: *The Road to Copenhagen* (2009)

Deforestation is a significant contributor to global greenhouse gas emissions

Figure 7
The Government's aim for climate change and energy policy

Decarbonise the UK		
and in doing so		
Keep our energy supplies safe and secure	Maximise economic opportunities	Protect the most vulnerable

The transition starts at home

The negotiations on a global deal are not a reason to delay action at home, but are a reason to act now. The UK is playing a leading role in driving international negotiations, and cannot expect others to sign up to a deal unless we show a firm commitment to change ourselves.

This Transition Plan shows how the UK is going to deliver the immense changes that are required. A transition of this scale needs a comprehensive plan to map the best opportunities, the sequences of decisions and the part that every sector, business and household can play, ensuring that the necessary changes are carried through in the most cost-effective way.

In delivering the transition to low carbon, this plan will also provide:

- security of our energy supplies;
- economic opportunities; and
- fairness for consumers, keeping bills as low as possible, particularly for the most vulnerable.

The rest of this document lays out the Transition Plan for the UK: a firm plan to 2020 including immediate actions, and analysis of the key pathways and scenarios to 2050.

By taking the lead internationally, the UK seeks to prevent dangerous climate change both at home and overseas

Chapter 2
Driving the transition

Summary

As set out in chapter 1, the scale of change we need in our economy and, in particular, our energy system is unparalleled. If we are to achieve this, Government will need to drive change by setting clear goals and a comprehensive plan to meet them.

Carbon budgets set the pathway. They are a world first: legally-binding caps on the greenhouse gases that the country produces over five-year periods. Informed by the evidence and monitored by the independent Committee on Climate Change, they chart the course to reducing greenhouse gas emissions by at least 80% below 1990 levels by 2050.

The first three carbon budgets were set in law following Budget 2009, committing to cut the UK's greenhouse gas emissions, compared to 1990 levels, by 22% in the current period, 28% in the period centred on 2015, and 34% in the period centred on 2020.

The UK has already made large savings. The country is set to deliver almost twice the greenhouse gas reductions the UK is committed to under Kyoto. But to stay within the carbon budgets, and have a contingency margin, a further 715 million tonnes of greenhouse gas savings are needed, including 459 million tonnes over the third period.

This will be stretching. It will need a view across sectors to make sure effort is focused in the most cost effective way, and one that covers both immediate action and long-term thinking: in short, a comprehensive plan.

The plan must harness the power of dynamic, competitive markets. However markets alone will not be enough - there is a strategic role for government. This means putting a price on carbon, driving new technologies, helping individuals and businesses to make informed choices, acting to maintain secure supplies (for example of the fossil fuels that we will continue to need for some time and in planning low carbon technologies), protecting the most vulnerable and maximising economic opportunities. Communities, too, have an active role: many are already organising themselves to bring about change in their own areas.

The plan also shows the balance of effort between sectors, based on bottom-up analysis of where savings are most cost-effective.

To stay on track, the Government is moving to a radical new approach. Every major decision now needs to take account of the impact on the carbon budget as well as the financial budget. The budget constraint will be real: failure to reduce emissions through policies and measures could mean a need to buy credits from abroad. A shortfall of 25 million tonnes of greenhouse gases, for example, assuming credits are £20/tonne, would mean a £500 million liability.

As important, the move to carbon budgets will mean that many more people have direct responsibility for action. The Government is piloting a new approach in which all major government departments have been allocated their own carbon budget, which will

Chart 1
The plan will reduce emissions in every sector

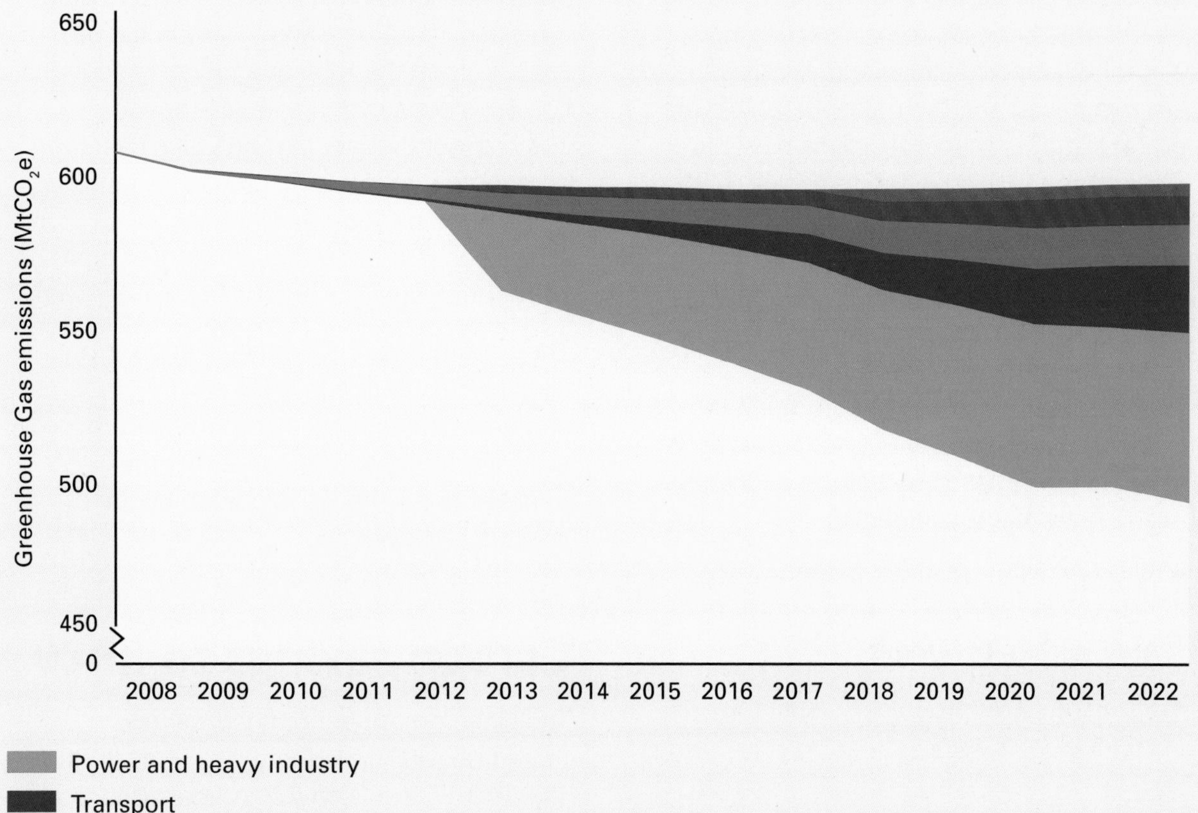

Source: Department of Energy and Climate Change

include emissions from their own estate and operations. In addition, delivery departments will have an allocation in respect of the areas of the economy they have policy influence over. Each department with a carbon budget will now produce its own carbon reduction plan by Spring 2010 to show how savings will be made.

This Transition Plan sets out the proposals and policies for meeting the UK's carbon budgets as required by the Climate Change Act.

Subsequent chapters lay out our Transition Plan sector-by-sector to show how carbon savings will be delivered.[1]

1. All announcements in this Transition Plan relating to support schemes are subject to State aid rules.

The previous chapter showed how demanding the task is: a rapid transition to a low-carbon economy, in doing so, ensuring that energy supplies remain secure, new economic opportunities are maximised, costs are minimised and the most vulnerable are protected. This chapter explains how the Government will drive the transition, with:

- A clear **pathway** to 2020 and beyond;
- A **comprehensive plan** to meet it: action will be achieved through dynamic and competitive markets, strategic government, and active communities; the plan will also set out how emissions reductions in each part of the economy add up to the total needed.
- A new system to share responsibility across Government to help the UK to **stay within its budget**.
- How the Government manages the costs of transition.

Setting the Pathway: carbon budgets

The UK's pathway is underpinned by the Climate Change Act, which became law in 2008. The Act commits the UK to achieving at least an 80% reduction in greenhouse gas emissions by 2050, compared to 1990 levels. This was the target recommended by the independent Committee on Climate Change as a fair and achievable share of the target to at least halve global emissions by 2050.

The Climate Change Act made the UK the first country in the world to adopt a long-term legal framework for reducing emissions: a system of five-year "carbon budgets", which provide a clear pathway for reducing emissions into the future. Carbon budgets are a limit on the total quantity of greenhouse gas emissions over a five year period. They reflect the fact that the UK's overall contribution to reducing global greenhouse gas emissions is determined by our emissions into the atmosphere over time, not by meeting specific targets in specific years.

The Committee on Climate Change also advises on the level of each budget, and provides scrutiny by reporting each year on progress. It ensures that the framework for the UK is guided by the science and evidence.

Carbon savings to 2020

The Government announced the first three budgets, covering the periods 2008-12, 2013-17 and 2018-22, in April 2009 (see table 1 below). These budgets are in line with those recommended by the Committee on Climate Change.

The UK is already on the path to cut emissions – and to deliver around twice the emissions reductions we are committed to under Kyoto – but the new carbon budgets will be stretching. The final budget period centred on 2020 requires a 34% cut on 1990 levels (or an 18% cut on 2008 levels), which requires an additional 420 million tonnes of savings.[3] This plan sets out the proposals and policies for meeting the UK's carbon budgets, as required by the Climate Change Act.

Chart 2 below provides an illustration of how the first three carbon budgets work.

2. The 2008 emission figures used throughout this Transition Plan are provisional.

Table 1
The first three carbon budgets

	Budget 1 (2008-2012)	Budget 2 (2013-2017)	Budget 3 (2018-2022)
Budget level ($MtCO_2e$)	3018	2782	2544
Percentage reduction below 1990 levels[3]	22%	28%	34%

Chart 2
The UK's carbon budgets are equivalent to a 34% cut in greenhouse gas emissions in 2020.[4]

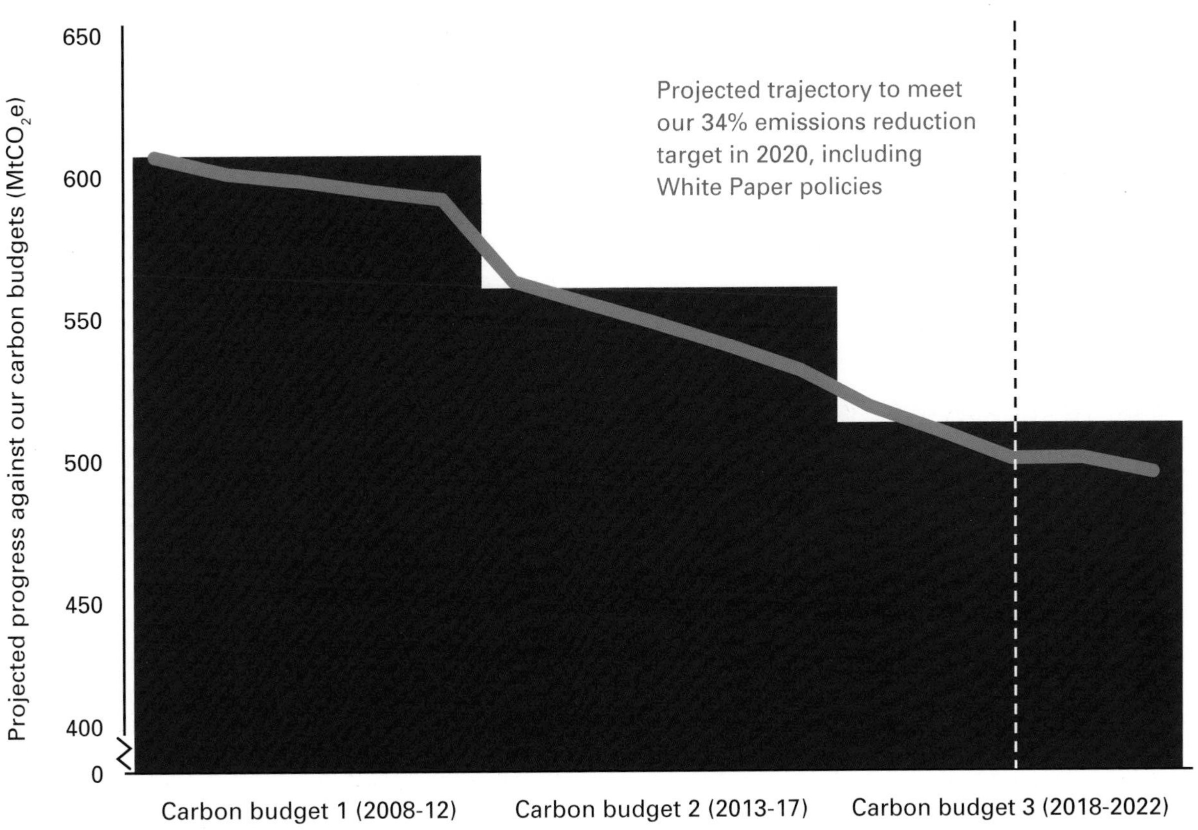

Source: Department of Energy and Climate Change

3. 1990 is the standard baseline for international comparisons, and the Climate Change Act 2008 requires that average annual emissions in the carbon budget period around the year 2020 (i.e. the third period, 2018-2022) are at least 34% below the 1990 baseline. This is referred to as a 34% reduction by 2020 for simplicity throughout this Transition Plan.
4. Comparing average annual emissions over the budget period to UK emissions in 1990 of 777.4 $MtCO_2e$.

Box 1
Climate Change Act: The First Three Carbon Budgets

The UK's carbon budgets and 2050 targets are set in law

What happens if an ambitious global deal is reached at Copenhagen?

The Government has committed to moving to tighter carbon budgets if a credible deal is reached at Copenhagen, and this will require the UK to cut its emissions even further. Tighter budgets are expected to be in line with the UK's share of the EU commitment to increase its 2020 target to as much as a 30% reduction on 1990 levels if other developed countries make comparable efforts under a new global agreement.

The extra effort to meet them can be made in different ways. As part of its 30% target, the EU will adopt a tighter cap for the EU Emissions Trading System (EU ETS). For emissions not capped by the EU ETS, the Government then has a choice between driving additional emissions reductions at home, for example through policies to incentivise cleaner transport or further efficiency improvements in housing stock, or purchasing offset credits that will deliver emissions reductions in developing economies. It will be guided by the most cost effective path towards the 2020 and 2050 targets, and will ask the Committee on Climate Change to review its recommended budgets following a global deal and once proposals on sharing out the new EU target are agreed.

Can offset credits from abroad be counted?

The carbon budgets set a limit on the level of the net UK carbon account. This is calculated by adjusting net UK greenhouse gas emissions to account for any carbon units which have been brought in from overseas by Government and others to offset UK emissions ('credits'), and UK carbon units which have been disposed of outside the UK ('debits'). The Government is aiming to meet the current budgets through domestic emissions reductions, without purchasing international offset credits, and this Transition Plan shows how this will be achieved. Delivering current carbon budgets through domestic effort will help prepare the UK for the move to a tougher 2020 target.

However, carbon units can still be bought or sold from abroad by companies participating in the EU Emissions Trading System, and the net UK carbon account must take account of this, whether it represents an overall credit against the budget (if, taken together, UK participants in the EU ETS buy more carbon units than they sell) or a debit (if they sell more than they buy).[5] The Government expects UK participants in the EU ETS to vary over the three budget periods between being net sellers and net purchasers of credits and allowances from abroad.

5. For further details of the carbon accounting system for carbon budgets, see http://www.decc.gov.uk/en/content/cms/what_we_do/lc_uk/carbon_budgets/carbon_budgets.aspx.

Do carbon budgets offer flexibility?

Lord Stern's Review of the *Economics of Climate Change (2006)* emphasises the importance of flexibility when designing policies to reduce the impact of climate change. The five-year budget system therefore allows "banking" and "borrowing" between budgets to increase the system's flexibility. Any emissions reductions which exceed those budgeted can be "banked" to help meet the next budget period, which rewards early action. Similarly, "borrowing" a limited quantity of emissions rights from the subsequent period will smooth out unexpected events towards the end of a budget e.g. a severe winter leading to higher energy demand and more emissions.

A comprehensive plan

Cost-effective changes require an economy-wide approach: effort must be concentrated where it will achieve the most for the least cost. To track delivery the Government must measure greenhouse gas savings across the economy to make sure overall emissions fall within the total budget.

Decisions cannot be guided by their short term impact alone. Changes to our homes and our energy infrastructure, for example, will have impacts lasting for decades so we must take action today while taking into account the changes that we will need to make in the future. As we move towards delivering a low carbon UK we must also act to maintain secure energy supplies, create jobs and ensure fairness. We therefore need a comprehensive plan.

A plan for governments, markets and communities

The plan includes a strategic role for government: putting the conditions in place for dynamic, competitive markets, mobilising communities and driving forward the transition to a low carbon economy and the sustainable development of the UK.

Government needs to put in place the necessary conditions for markets to be competitive. This means an independent regulator with responsibility for ensuring competition and proactively protecting consumers interests.

But markets alone cannot deliver a transition of this speed or scale. Failures in the market mean people or businesses can take decisions that are right individually, but lead to worse outcomes for society as a whole.

Lord Stern's Review of the Economics of Climate Change (2006) showed that in reducing emissions governments need to act in three ways to deliver change. These broadly guide the Government's strategic interventions.

First, governments should put a price on carbon emissions to reflect the true costs of the damage caused through climate change, and to give incentives to consumers and companies to move to cleaner technologies or to change their behaviour. Chapter 3 outlines how the power sector and heavy industry now have to factor in the price of carbon through the European cap-and-trade scheme, the European Union Emissions Trading System (EU ETS).

Second, government must drive the new technology and infrastructure that is needed. Businesses do not invest as much or as quickly in these areas as society needs, both because of the risks involved and because many of the benefits of investment do not accrue to the firm, but to society more generally. For example the Government is funding incentives and infrastructure to encourage swifter deployment of electric cars.

Third, government must help people make low carbon choices. People and businesses often face obstacles such as upfront costs, a lack of information, or inertia. Government provides information and wider support for households to install insulation.

To maintain energy security, the Government needs to ensure a supportive investment climate so that new low carbon energy infrastructure is built and, because we will be reliant on fossil fuels for many years to come, to ensure that we maximise the UK's own economic reserves of oil and gas and have secure supply chains for the energy we do import.

To maximise the economic opportunities of transition, the Government is building up the skills and infrastructure needed, developing regional hubs and putting in place support for the new industries of the future, as described in chapter 5 and in the *UK Low Carbon Industrial Strategy* published in par allel with this Transition Plan. And the Government is acting to secure fairness, through providing support to the most vulnerable and improving the energy efficiency of their homes.

Tackling climate change is clearly crucial to protect our natural environment. But the Government will work to ensure that the action it takes improves the environment more widely, making our cities healthier and quieter, cutting our resource use and protecting the natural environment. But where impacts cannot reasonably be avoided, the Government will take steps to minimise them.

The UK has a target to produce 15% of its energy from renewable sources by 2020

Chart 3
The UK Transition Plan will cut emissions by 459 million tonnes in the third carbon budget (2018-2022).

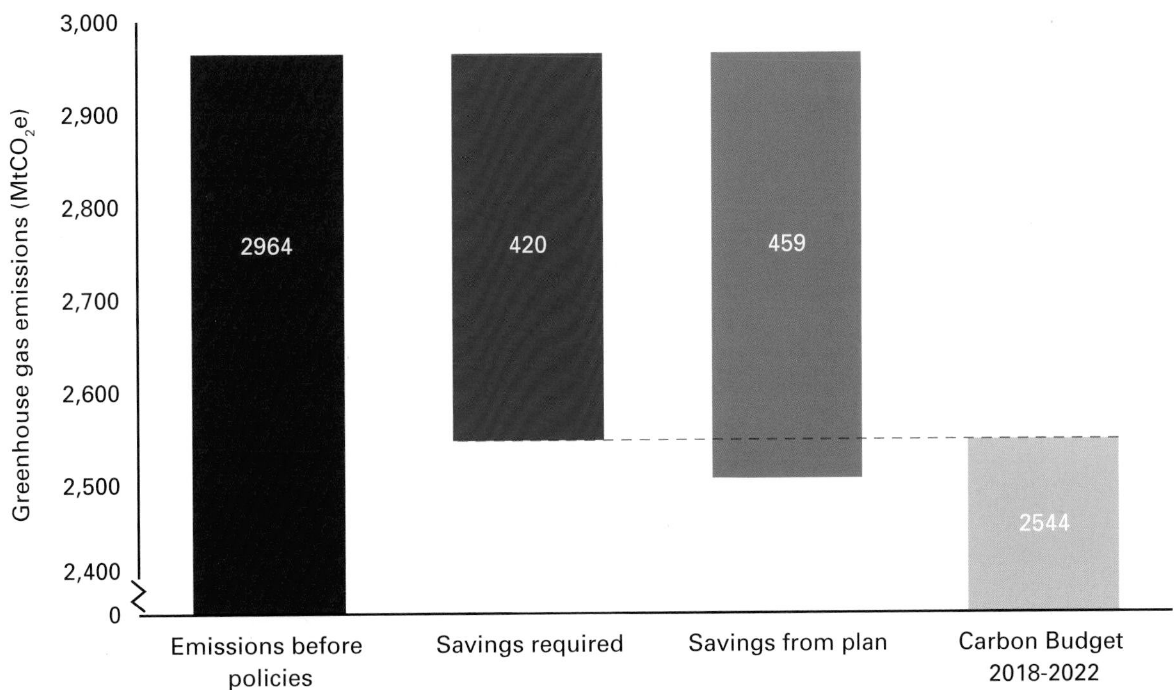

Note: Savings also include interaction effects. Refer to Table A1 for a full breakdown of carbon savings in the third budget period
Source: Department of Energy and Climate Change

The balance of effort

A comprehensive plan also needs to include communities, individuals and businesses. Many thousands of people are already working together, exchanging ideas and spurring each other on. The transition cannot happen without them, and this plan announces a range of further support (see chapter 4).

The UK's Transition Plan sets out how we will cut emissions by 459 million tonnes by the third carbon budget. On the basis of current central projections the Government will over-deliver the reductions needed to meet the third budget by a margin of 39 million tonnes in that period, and by 147 million tonnes over all three carbon budgets. This will help provide a buffer against uncertainty, and prepare for a tightening of the carbon budgets that will follow an ambitious international deal.

This will need action in every sector of the economy. Chart 4 shows where the savings will be made, based on the policies in succeeding chapters.

A major contribution to meeting the carbon budgets comes from the transition to renewable sources of energy, covering electricity, heat and transport. Under an agreement to drive the uptake of renewable energy across Europe, 15% of energy in the UK must be renewable by 2020 (see chart 5). The UK Renewable Energy Strategy is published in parallel with this document.

Chart 4

The main policies driving emission reductions are the EU Emissions Trading System, energy efficiency policies, and increased use of renewable energy for heat and transport

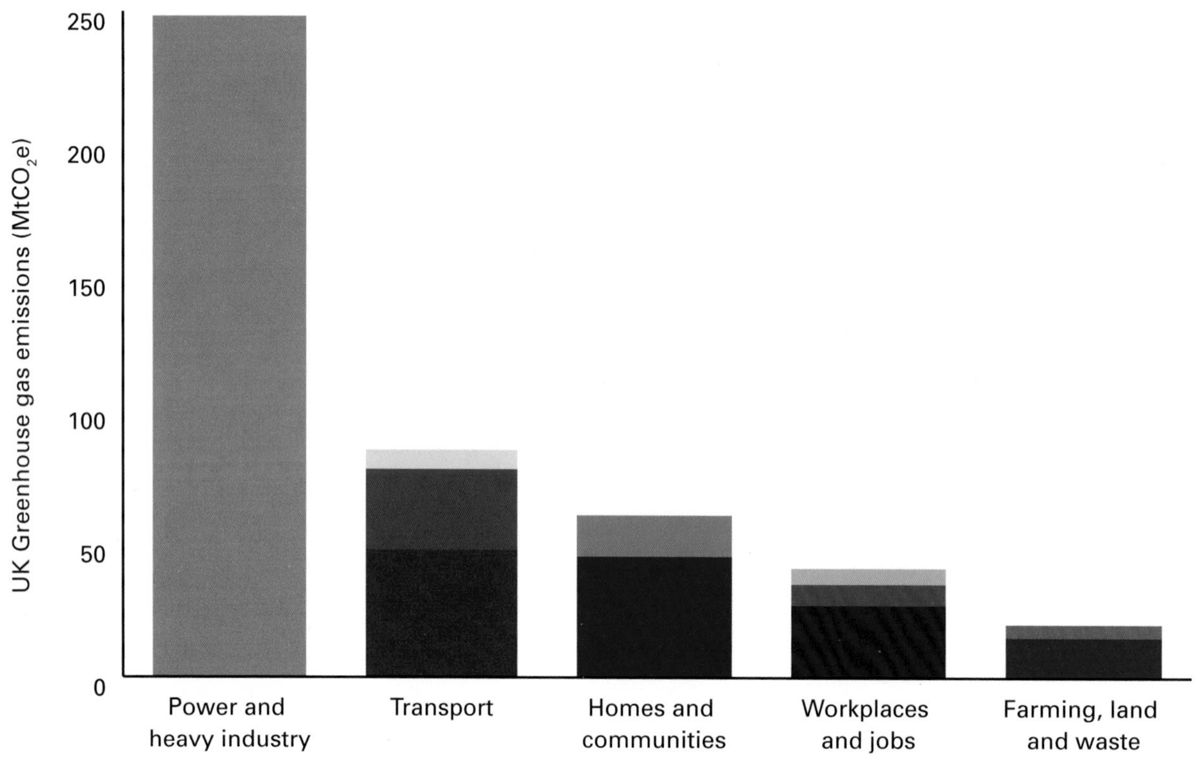

- European Union Emissions Trading System
- New vehicle CO_2 policies
- Additional renewable transport fuels
- Low carbon buses, car improvement technologies, driver training, illustrative rail electrification of 750km of track
- Energy efficiency, smart metering, Community Energy Saving Programme, and zero carbon homes
- Clean energy cashback (renewable heat incentive)
- Clean energy cashback (renewable heat incentive)
- Climate Change Agreements and other policies
- Carbon Reduction Commitment and other policies
- Farming (crop management, manure management etc.)
- Waste policies (diverting waste from landfill, increased landfill tax)

Source: Department of Energy and Climate Change

Chart 5
Renewable energy use will grow seven-fold in the next decade

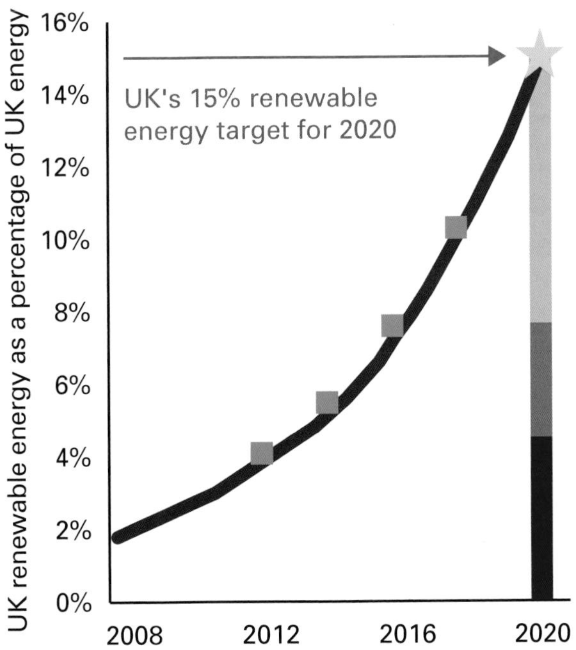

Source: Department of Energy and Climate Change

Staying on budget

These budgets need to be met despite uncertainty about the future. Some policies, like the EU Emissions Trading System, guarantee that net emissions will be no higher than the limit or 'cap' that is set. However, in other sectors the Government cannot be completely confident in advance about future emissions. There is uncertainty about what policy will deliver and whether other factors, such as faster than expected economic growth will drive up emissions.

Chart 6 shows that the UK is on track to meet its carbon budgets on central projections and has factored in an additional contingency margin to mitigate some of the uncertainty described above.

There are a number of other ways in which the Government is working to increase the certainty that the UK will meet its carbon budgets.[6] Emissions projections have been improved to remove any double counting of carbon savings, and the Government is continuing to explore new policy options, such as new ways to help small businesses to save carbon.

The introduction of carbon budgets introduces a new imperative: they are legally binding and must be met. As a result there are also risk management tools that could be used if required as a last resort: the provision to "bank" and "borrow" (within defined limits) between budget periods, and to buy international credits to offset domestic emissions. In effect, there will be a cash penalty for failing to meet the plan: a shortfall of 25 million tonnes of greenhouse gases, for example, assuming credits are £20/tonne, would mean a liability of £500 million. While they remain an insurance option, these tools are not part of this Transition Plan, and the Government has set a legal limit preventing it from buying credits for the first budget period.

There is uncertainty about future emissions, but the Government is working to ensure that carbon budgets are met

6. Further details are provided in the Analytical Annex published alongside this Transition Plan.

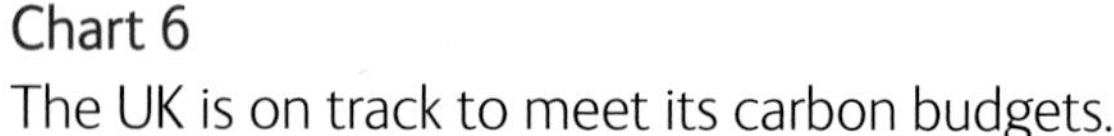

Chart 6

The UK is on track to meet its carbon budgets.

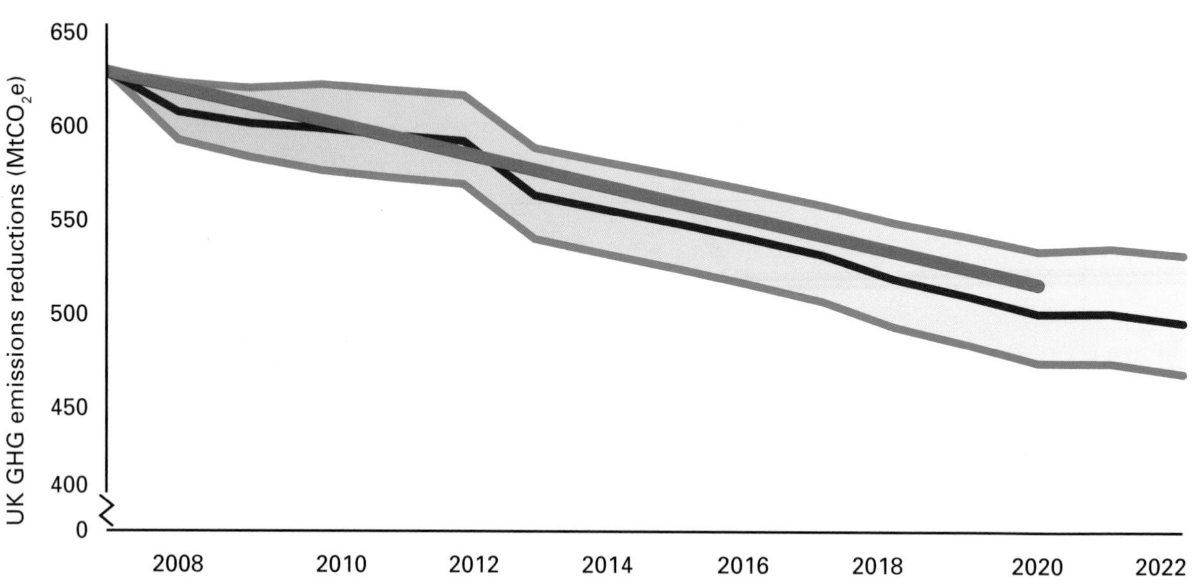

- Central emission projections
- Upper and lower emission projections
- illustrative straight line trajectory to meet the carbon budgets and achieve a 34% reduction in emissions by 2020

Annex A shows the "indicative annual range" for the net UK carbon account over the first three carbon budgets, the Government's expectation of the range over which the net UK carbon account might fall, taking into account uncertainty.

Taking responsibility across government

Every part of Government will need to help drive the transition that is needed to live within the UK's carbon budgets. For the first time, each major government department will now have its own carbon budget representing its share of responsibility (see chart 7). This pilots a new approach in Government.

The budgets will be made up of two elements.

First, an allocation is made to reflect emissions from Departments' own estate and operations. *From April 2010, it will also include emissions from schools, further and higher education institutions and the NHS.*

Second, an allocation is made to all delivery departments depending on their degree of influence on reducing emissions in each sector of the economy. This will incentivise departments to work together in a collaborative way to reduce emissions across the economy. For example, the Department for Transport and the Department for Children, Schools and Families will have an interest in helping parents to make choices that will reduce emissions from the school run, such as using yellow school buses or cycling to school.

The Government's plans are for emissions reductions in the government estate to be at least as ambitious as the rest of the economy. *For the Department of Energy and Climate Change (DECC) to set an example, it has challenged itself to reduce its own emissions by 10% in 2009 -10.*

To underpin the delivery of its carbon budget, each department will publish its own carbon reduction plan by Spring 2010. These will set out in detail the actions the department will take, on its own and working with others, to reduce greenhouse gas emissions in the parts of the wider economy that it can influence, as well as from its own estate and use of transport. These plans will include milestones and indicators to measure progress and ensure that planned actions are delivered.

Every department will also publish its plans to cope with the effects of climate change that are already unavoidable. The details of the budget-setting process and the budget for departments in each period are given in Annex B. This approach will be reviewed ahead of the second budget period.

HM Treasury will play a key role in the departmental carbon budget system, supporting the delivery of carbon budgets in keeping with its role at the centre of government.

The UK Government is one of the first to integrate climate change and energy issues in one department, through the creation of the Department of Energy and Climate Change (DECC). DECC has overall responsibility for ensuring delivery of all the measures set out in this plan. Following publication, and in parallel with the emerging work on the vision for 2050 (set out in chapter 8), *DECC will work with its delivery partners and wider stakeholders on the necessary arrangements to ensure the commitments set out here today are delivered effectively.*

Chart 7

The percentage share of carbon budgets in 2018-2022 from each department

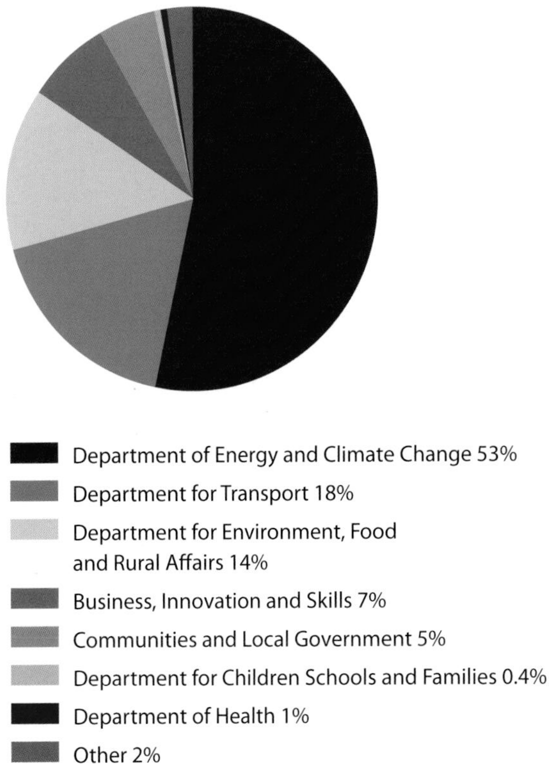

Source: Department of Energy and Climate Change (2009)

Managing costs

As described in chapter 1, tackling climate change is the lower cost option for Britain: failure to act would mean more extreme droughts and floods, greater dependency on imported fossil fuels, and a missed opportunity to lead new low-carbon industries.

The UK can meet its carbon budgets at the same time as doubling the size of its economy by 2020 on 1990 levels,[7] and in doing so deliver extra benefits, including improved energy security and local environmental quality. The UK is expected to rely on net imports to meet around 45%

of its net gas demand in 2020, compared to the level of around 60% expected without the Government's policies to reduce demand (see chart 8).

There is therefore a compelling argument for pursuing the policies set out in this Plan. They do come at a cost however – such as extra investment in low carbon technologies, which needs to be properly managed.

Table 2 presents today's value of the costs of the policies set out in this plan including both public and private costs.[8] Total net costs over the lifetimes of these policies are estimated to lie between £25 and £29 billion.

These net costs are significant but are broadly consistent with other estimates of the costs of action on climate change.[9] Net cost estimates for some components of the package are somewhat lower than previously estimated owing to changes in fossil fuel price assumptions and wider factors.

In developing the package of policies, the Government has focussed wherever possible on the most cost effective sources of emissions reductions. As highlighted in table 2, many of the policies to meet carbon budgets have a negative cost, which means they help tackle climate change and save resources for the economy while saving on bills. This is particularly the case for policies to improve energy efficiency in the household sector (see chapter 4). As set out in chapter 5, the transition to low carbon will bring costs but also business benefits.

The impact from the policies to households and businesses will be through higher prices for some carbon-intensive goods and services and changing patterns of consumption. However, the most significant impact on consumers will be an increase in energy bills. The additional impact in 2020 of the policies in this plan is equivalent to a 6% increase on current household energy bills and a 15% increase on current business energy bills.

The Department of Energy and Climate Change will work with its delivery partners and wider stakeholders to ensure the commitments set out here are delivered effectively

7. Based on an assumed GDP growth rate of 2.25% per year.
8. The costs can be judged to be consistent with the overall costs of delivering the long term target that were estimated in the Climate Change Act Impact Assessment of £324 to £404 billion.
9. The overall costs of delivering the 2050 target were estimated in the Climate Change Act Impact Assessment to be in the range £324 to £404 billion

Table 2
The lifetime net costs of this plan

Sector	Lifetime net cost in today's terms (£ billion)
Power and heavy industry	48.7 to 53.0
Transport	6.4
Workplaces and jobs	-2.9
Homes and communities	-26.9 to -27.9
Farming, land and waste[10]	0.1
Total	**25.4 to 28.7**

Chart 8
The transition will help reduce dependence on imported gas

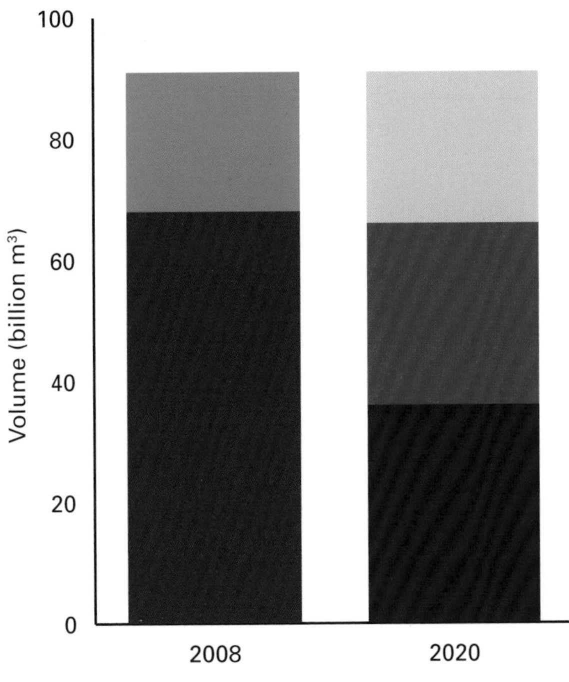

Source: Department of Energy and Climate Change

These figures do not take account of the long-term benefits of British leadership in new industries and services, or the long-term costs of failing to tackle climate change.

The Government continues to seek to minimise the upward pressure on prices through policies to help households and businesses improve their energy efficiency. This is set out in chapters 4 and 5.

To ensure transparency, accountability and value for money in costs associated with tackling climate change, the Department for Energy and Climate Change will report on the financial flows to tackle climate change through instruments which impose costs on energy bills, as required for National Accounts purposes. To ensure long-term certainty for investors while acknowledging the uncertainties associated with future technology costs and deployment rates, DECC will also undertake regular reviews of support levels across the range of levy mechanisms to support climate policy.

Chapters 3-7 look sector by sector to explain how carbon savings will be achieved and in doing so support security of energy supplies, economic opportunities and protect the most vulnerable.

10. Costs shown relate to waste proposals. Farming has negative resource cost; policy costs not yet available.

Chapter 3
Transforming our power sector

Summary

Currently three quarters of our electricity is generated using coal and gas. By 2050 we may need to produce more electricity than we do today but must do so largely without emitting greenhouse gases. So we will need to transform our system so that electricity is generated from clean sources such as renewables, nuclear and fossil fuel plants fitted with carbon capture and storage technology. To support these changes, we will need an electricity grid with larger capacity and the ability to manage greater fluctuations in electricity demand and supply. To make this transition, the Government needs to maintain the right conditions for energy companies to invest very large sums in new power stations and in the transmission and distribution networks.

This Transition Plan, along with wider policies, will put us well on this path, with around 40% of our electricity coming from these low carbon sources by 2020. It will cut emissions from the power sector and heavy industry by 22% on 2008 levels by 2020.

At the heart of the Plan is the EU Emissions Trading System which sets a declining limit or 'cap' for emissions.

Chart 1

The power and heavy industry sector will contribute over half of the additional savings in 2018-22

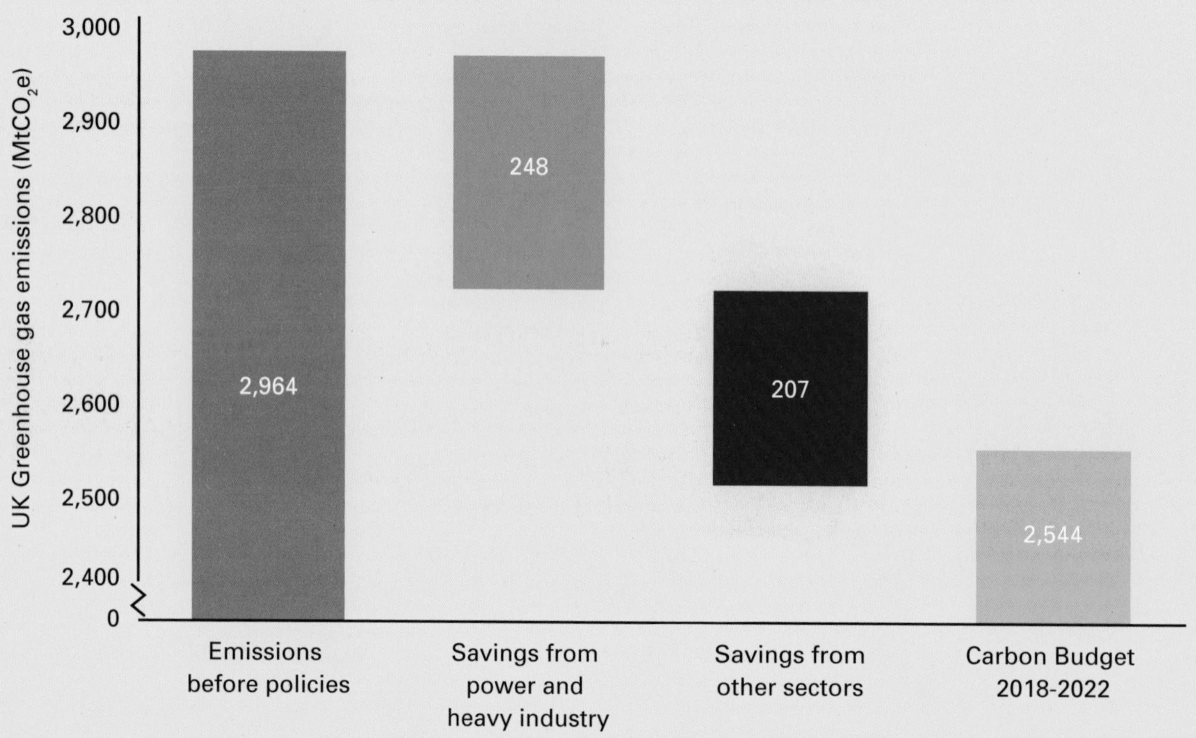

Note: Reductions due to policies introduced prior to the Energy White Paper 2007 are not shown.

Over 2008-12 the UK's annual cap under EU ETS has been reduced by 30$MtCO_2$ compared on like-for-like basis with the cap in 2005-7.

Source: Department of Energy and Climate Change

But this alone will not be enough to enable the rapid development and use of low carbon technologies. So the Government is taking further action:

- Renewables: the Government is going further to support renewable technology by increasing financial incentives for renewables developers through the Renewables Obligation and the forthcoming clean energy cashback initiative. It is also streamlining the planning process, supporting innovation, speeding up grid connection and developing UK supply chains. To provide targeted support to UK businesses the Government is launching the Office for Renewable Energy Deployment. In addition, the Government is publishing a short list of Severn Tidal projects for further study.
- Nuclear: the Government's Office for Nuclear Development has made strong progress in facilitating the building of new nuclear power stations by energy companies, by taking action to streamline the planning and regulatory approvals processes for new nuclear power stations. The Government is currently assessing sites nominated by potential developers, to establish which are potentially suitable for the deployment of new nuclear power stations by the end of 2025. This assessment will be included in a draft National Policy Statement for nuclear power, which the Government will publish for consultation later in 2009.
- Carbon capture and storage: in 2007 the Government launched a competition to build one of the first commercial scale carbon capture and storage demonstration projects in the world. In April 2009 the Government announced that new fossil fuel power stations would have to be designed and built so that they could fit CCS in the future. In a consultation launched in June 2009, the Government proposed a new financial and regulatory framework to drive the development of CCS. These proposals included plans to fund up to four CCS demonstrations in the UK and a requirement for any new coal power station to demonstrate CCS. The Department of Energy and Climate Change will also establish an Office of Carbon Capture and Storage to support the delivery of this work.

The Government is working with the regulator (Ofgem) and industry to increase grid capacity and to support development of new technologies which could enable the grid to work better in the future. The Government will later this year publish a high level vision setting out what a future smart grid could look like, linked to wider work on the roadmap to 2050 (see chapter 8).

The Government would like to see a smooth transition from the existing system to a new cleaner power sector. It expects the risks to security of electricity supply to be manageable over the next decade as new investment comes through to replace closing power stations. In the longer term, the Government will need to ensure that it maintains security of supply as low carbon technologies become increasingly important. Government plans to issue a call for evidence later in 2009 to seek views on these issues.

The scale of the challenge

Three quarters of the electricity we use in our homes, businesses and public buildings is produced from coal and gas (as shown in chart 2). The way in which electricity is produced and supplied to us is described in box 1. But by 2050 the UK will need to produce very few greenhouse gas emissions overall. As set out in chapter 8, taking into account the costs and potential of all the options it is likely that we will need to reduce the emissions from the power sector to almost zero.

To do this, we need to produce our electricity from low carbon technologies such as renewables, nuclear and fossil fuel fired generation fitted with carbon capture and storage technology. We also need a bigger, smarter electricity grid that is able to manage a more complex system of electricity supply and demand.

Chart 2

Around 75% of our electricity is currently generated from gas and coal today; renewables will expand to around 30% of our generation by 2020[1]

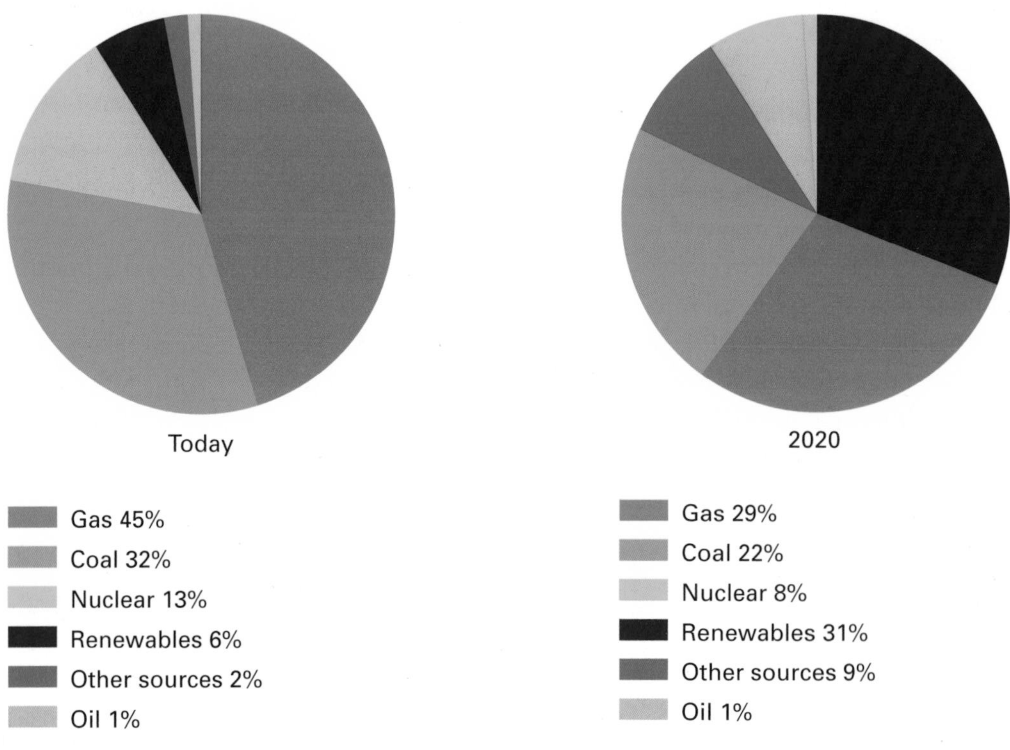

Source: Energy Trends (2009,Quarterly)Department for Business Innovation and Skills (2009)

Source: Department of Energy and Climate Change

1. The 2020 chart is a projection of possible shares of electricity generated from different sources, from the DECC energy model which assumes existing nuclear power stations are closed in line with published retirement dates and 1.6GW of new nuclear capacity is constructed by 2020. Estimated energy demand today and in 2020 is around 370TWh.

Box 1
Electricity production and supply

Electricity is generated by privately owned companies operating in a competitive market within the policy framework set out by the Government. The Office of Gas and Electricity Markets (Ofgem) has responsibility for ensuring effective competition and enforcing regulation in the energy market while the Government's policy framework is set to achieve public policy goals such as tackling climate change.

Most electricity is produced in power stations which burn fossil fuels (coal, oil or natural gas) or use nuclear energy. A small, but increasing amount of our electricity is generated in other places such as wind farms.

Three Transmission Owners (National Grid in England and Wales and SP Transmission Limited and Scottish Hydro-Electric Transmission Limited in Scotland) and 14 Distribution Network Operators construct and operate the electricity grid under licence from Ofgem which allows electricity to flow from where it is produced to where it is needed. National Grid also acts as 'system operator' ensuring the amount of electricity produced and consumed is balanced second-by-second. Consumers then buy their electricity from electricity suppliers; these are also private companies.

The plan to 2020

The policies set out in this Transition Plan will help ensure we cut emissions from electricity and heavy industry by 22% on 2008 levels by 2020, and secure our electricity supplies. The private sector will be responsible for bringing forward the investment needed to deliver this change, but because of the importance of the task and the public policy choices involved, the Government will need to lead this change by setting out clear goals and an appropriate policy framework. The Government's approach to decarbonising our electricity system is to apply a carbon price through the EU Emissions Trading System (EU ETS) (set out in more detail below), and to support the rapid development and use of low carbon technologies. Already, electricity generated from renewables has more than doubled in the last five years, and action is being taken to facilitate new nuclear power and the demonstration of carbon capture and storage, to reform the planning system and to review the process of connecting new generation to the electricity grid.

But Government and private sector must together focus on delivery. The Department of Energy and Climate Change (DECC), has already piloted a new approach to delivering more proactive support to industry in the Office for Nuclear Development (OND), and the Government will be pursuing the same approach for renewables and carbon capture and storage, by setting up an Office for Renewable Energy Deployment (ORED) and an Office of Carbon Capture and Storage; all aim to remove barriers to vital investment, and make use of both private sector and regulatory expertise in carrying out their work.

The extent to which companies will build new power stations ultimately depends on the expected profitability of such investments which relies on the way in which they believe factors such as fossil fuel, carbon and electricity prices, technology costs and regulatory or planning risks will evolve over time.

So in taking action, the Government remains mindful of the need to provide as much clarity as possible in the market framework to enable private sector investment to come forward. The Department's plan involves:

- Incentivising the power sector to **reduce emissions through the EU ETS**
- Taking action to make sure **renewables, nuclear and carbon capture and storage** can play their part
- Putting the measures in place to create **a bigger, smarter grid**, ensuring fair and timely access
- Using electricity more **efficiently**
- Keeping our **electricity supplies safe and secure**.

Chart 3

The EU Emissions Trading System will deliver significant emission savings between 2008-2022

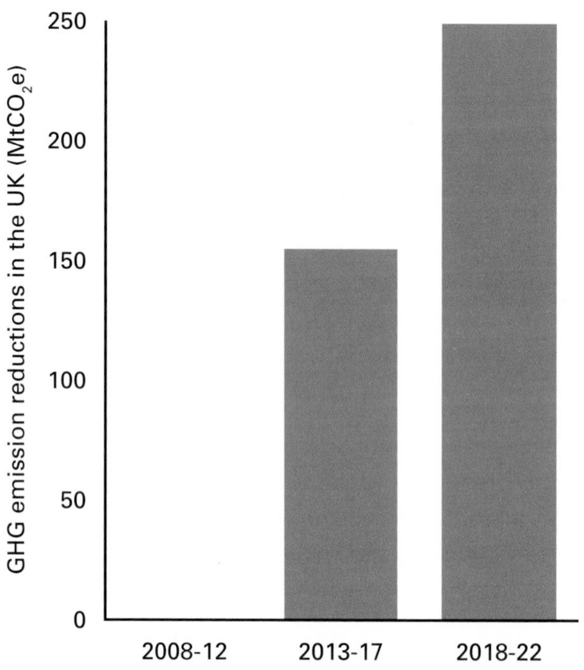

Note: Reductions due to policies introduced prior to the Energy White Paper 2007 are not shown.

Over 2008-12, the UK's annual cap under EU ETS has been reduced by 30$MtCO_2$ compared on like-for-like basis with the cap in 2005-7.

Source: Department of Energy and Climate Change

Putting the plan into practice

Incentivising the power sector to reduce emissions

The best way of incentivising the most cost effective mix of low carbon technologies is to put a limit or 'cap' on emissions. Since 2005, the European Union Emissions Trading System (EU ETS) has set a declining cap on emissions from the large industrial sectors, including power.

The EU ETS is the first multilateral carbon trading system of its scale, anywhere in the world, and is expected to account for over 65% of the emissions savings in Europe by 2020.[2] It will reduce Europe's emissions by around 500 million tonnes in 2020, which is about the same as the UK's carbon dioxide emissions this year.

2. Other countries, such as the United States and Australia, are developing similar 'cap and trade' systems, which may link to the EU ETS in the future, making it a cornerstone of a global carbon trading system.

The EU ETS is the single most important policy to reduce UK emissions (covering about half of the UK's carbon dioxide emissions) and is expected to deliver emissions reductions from the power sector and heavy industry of 22% on 2008 levels by 2020 – underpinning the transition to low carbon electricity generation (see chart 3).

It tackles emissions from large sources such as electricity generation and heavy industry;[3] it also allows the companies involved to trade the right to emit with each other, creating a carbon price and enabling emissions cuts to be made where they are cheapest. As the carbon price generated by the EU ETS increases, it makes producing electricity from high carbon power stations less and less attractive and creates an incentive for power station operators to invest in cleaner electricity generation (see box 2).

The EU ETS and other policies to bring forward low carbon technologies – set out below – put us on track to decarbonise the UK's electricity system by around 40% by 2020. There will need to be significant further change in the period beyond 2020. Chapter 8 (2050) sets out the possible trajectories for reducing power sector emissions further. The Government is planning for an ambitious global deal to reduce global emissions at Copenhagen, which will mean the EU acting to reduce its emissions by 30% by 2020. This will lead to a significant tightening of the EU ETS cap and a system where all companies pay for their emissions and a guarantee that most of these reductions take place in the EU.

Box 2
What is a 'carbon trading system'?

A carbon trading system, like EU ETS, places a cap on the total emissions of participants, divides that cap into rights to emit or 'allowances', and allows participants to trade those allowances.

Each participant must surrender allowances proportional to the number of tonnes of carbon dioxide they emit. This encourages them to reduce their emissions if they can do so for less than the cost of allowances. Allowances are initially distributed either freely to participants, or through an auction. Once distributed allowances can be traded between participants. The trading of allowances gives them a value, or price, which is the same for all participants. This common price creates the same incentive for all participants to reduce their emissions which keeps the costs down by incentivising the least cost options to reduce emissions across the whole of the system.

3. From 2012 aviation emissions will also be included in the EU ETS (see chapter 6).

Introducing a carbon price as just described is critical to achieving clean electricity generation but it is not the complete answer. This is because other barriers exist to the development and deployment of innovative technologies such as high development costs, uncertainties around volatile fuel prices and technology costs and the lack of an effective supply chain.

The Government looks forward to the first UK progress report from the Committee on Climate Change in October 2009, which will include details on the capability of the power sector in delivering decarbonisation at the pace required.

Taking action to bring forward renewables, nuclear and carbon capture and storage

There are some barriers which are common to all electricity generation technologies and the Government is taking action to remove them. However other barriers that need to be removed are specific to each technology or group of technologies. Renewable electricity, nuclear and carbon capture and storage will be needed in some combination, and the Government is therefore taking action to enable each one to contribute as part of the mix.

Box 3
New Single Consent Regime for Infrastructure and Infrastructure Planning Commission

The Planning Act 2008 provides for a new independent Infrastructure Planning Commission to take decisions on nationally-significant energy infrastructure projects, and this will happen from 2010.

Following public consultation and Parliamentary scrutiny, Ministers will designate National Policy Statements (NPSs) for infrastructure development. These will set out the national need for energy infrastructure and other guidance on national policy that the IPC needs to consider when making decisions. Planning authorities, including responsible regional authorities preparing Regional Strategies, must have regard to these new NPSs when preparing development plans and, where relevant, when making planning decisions under the Town and Country Planning system. This will help ensure that decisions on renewables and other sectors, whether large or small, are taken consistently. The Government will publish the first of these NPSs for consultation later this year.

The new regime will provide three opportunities for communities and interest groups to have their say: during the public consultation on the NPSs; through local consultation at the pre-application stage; and, by making representations to the IPC when it considers an application.

Supporting all generation technologies

The Government is undertaking major planning reform to ensure that planning supports the deployment of all electricity technologies (see box 3)

These changes will create greater clarity and predictability in the planning framework for nationally significant energy infrastructure and better opportunities for the public to engage with the process, with increased transparency and accountability for decisions.

The Government is also taking action to help ensure that high-quality forward-looking information is available. Investments in new infrastructure will be informed by views of future changes in energy prices and demand. Through the Energy Markets Outlook, the Government provides information that potential investors may find helpful in taking decisions on energy infrastructure.

The Government is improving the way it supports technology innovation which will affect all low carbon power technologies. Further details are set out in chapter 5.

Renewable power

Delivering large increases in renewable electricity will be critical in decarbonising the power sector. However generating electricity using renewables still costs more than fossil fuel generation and deployment is hampered by a range of barriers such as access to transmission capacity and lack of an effective supply chain.

The Government has already made progress in removing these barriers and is publishing the UK's Renewable Energy Strategy (2009), alongside this document. This sets out how the UK will increase its energy from renewable sources almost seven-fold, and the role of the electricity sector in contributing to this.

Box 4
The Renewables Obligation provides financial incentives for renewable electricity generation

The Renewable Obligation provides a financial incentive to invest in renewables by placing an obligation on electricity suppliers to source a certain proportion of the electricity they sell to customers from renewable sources.

Renewable energy generators receive Certificates, known as Renewables Obligation Certificates (ROCs) for the renewable electricity they produce, and can then sell these to electricity suppliers, who use them to meet their obligations.

They demonstrate this by submitting the ROCs they have bought. If they are unable to present ROCs for the whole of the specified amount of electricity, they have to pay a penalty. These payments are redistributed to suppliers who did present ROCs. It is this redistribution that provides the incentive for suppliers to present ROCs rather than simply paying the buy-out price.

ROCs can be sold with or without the electricity they represent, meaning that they provide generators with financial support above what they receive from selling their electricity in the wholesale market. Different technologies receive different numbers of ROCs, to account for differences in technology costs.

Around 30% of our electricity is expected to come from renewables by 2020; highlights of how this will be achieved are outlined below.

Ensuring large scale renewable power makes financial sense

To ensure that investing in renewables makes financial sense for investors and help bring down the costs of renewables in the future, in 2002 the Government introduced the Renewables Obligation (RO) (see box 4 above).

Since the RO was introduced, renewable electricity generation has tripled from less than 1.8% in 2002 to around 5.3% in 2008. The UK is now number one in the world for installed offshore wind capacity. The wind farms under construction today and those awaiting construction will together produce enough electricity for another 5.5 million UK homes.

Chart 4

Renewable electricity has tripled since the introduction of the Renewables Obligation

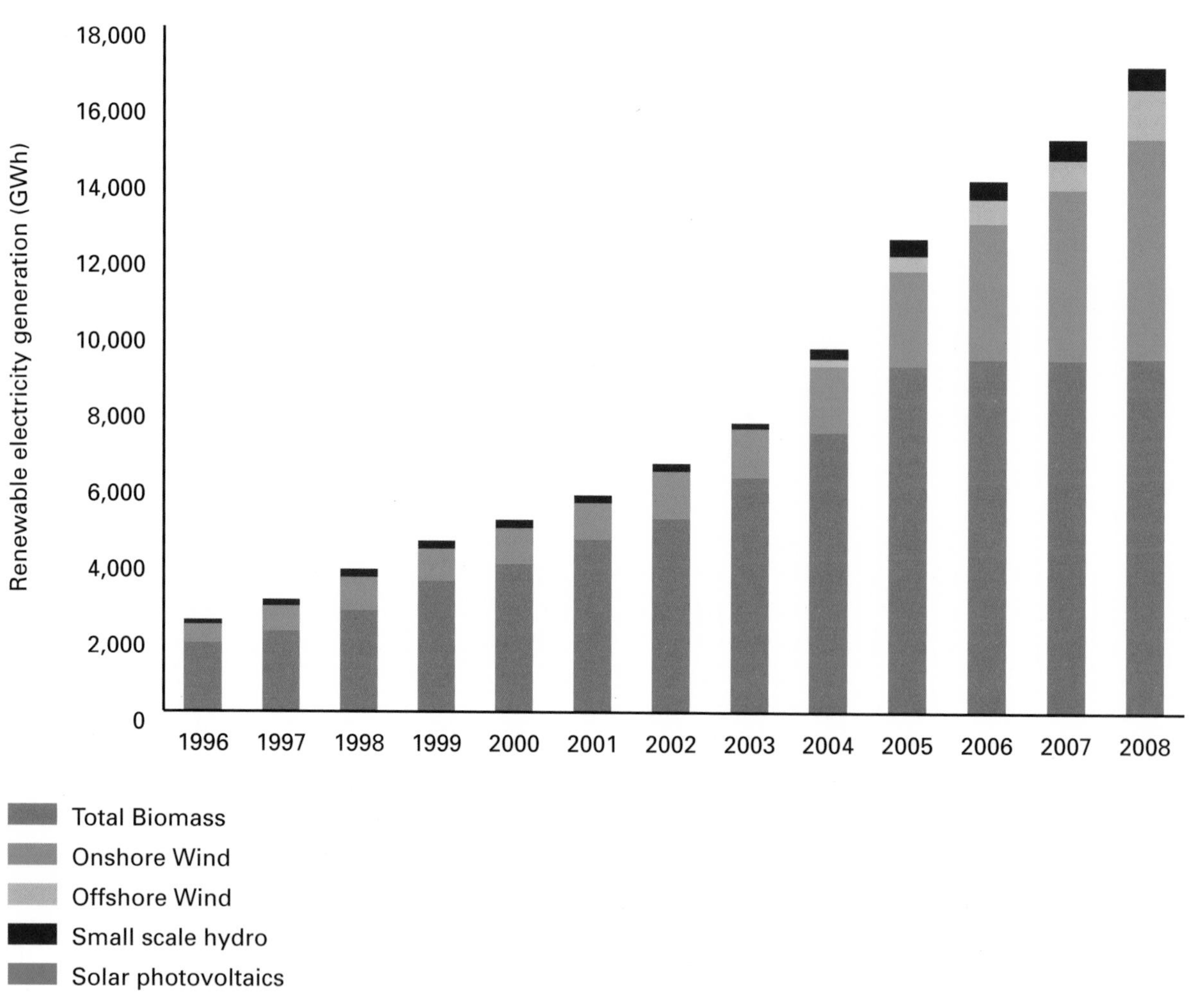

Source: Digest of UK Energy Statistics; Energy Trends (2009)

By 2010 the RO along with exemption from the Climate Change Levy (see chapter 5) will be worth around £1 billion a year to the renewable electricity industry. However we need to do more if the UK is to further increase the levels of renewable electricity:

- In November 2008, the Government confirmed that the RO will be retained and extended from 2027 to at least 2037, therefore giving certainty to investors about long-term support for investment in renewable generation.[4]
- In April 2009, the Government announced that it would be reviewing the level of support to offshore wind to determine whether it should be increased. In total it expects up to 3 GW of planned projects could benefit from any increase in support – enough to power an estimated 2 million homes.

Improving grid connection

The Government is delivering a new strategic approach to investment in the grid both onshore and offshore, and speeding up access arrangements, so that renewable, new nuclear and other low carbon generation can connect to the grid when it needs (further detail set out below).

Ensuring planning supports renewables deployment

As set out above, the planning system will need to play a central role in supporting the deployment of renewable energy. Improvements to the system are already being made with the provisions in the Planning Act 2008, which creates a new system of development consent for nationally significant infrastructure projects with the Infrastructure Planning Commission taking decisions on larger renewable projects including large wind-farms in England and Wales (over 100 MW offshore or 50 MW onshore).

But further action is needed at regional and local level to ensure projects below these thresholds are also supported whilst ensuring development takes place in appropriate places, at the right time, and in a way that gives business the confidence to invest. The UK Renewable Energy Strategy (2009) also sets out reforms to the Town and Country Planning system which are designed to deliver a more effective, transparent and responsive system in England. This includes a package of measures designed to support more effective and proactive planning by local and regional authorities so that they are better able to capitalise on the renewable opportunities available to them.

Of course, there will always be locations where new infrastructure will not be appropriate and the Government will continue to apply higher planning tests in National Parks and Areas of Outstanding Natural Beauty.

Developing the renewables supply chain

The Government wants to maximise the economic and employment opportunities for Britain in expanding renewable sources of energy, putting us at the forefront of global competition in the low carbon economy. The UK must make the most of its strengths as one of the world's largest manufacturing economies, as a world class centre of expertise and as a leading location for inward investment. The increase in deployment of renewables both here and in other countries has highlighted the bottlenecks and constraints in the supply chain. The Government needs to ensure that the renewables industry and its supply chain can deliver the unparalleled deployment required. Further rapid growth will depend on more investment in innovation and manufacturing, the availability of skills, and the development of infrastructure such as ports and construction vessels for offshore deployment.

4 The consultation on Renewable Electricity Financial Incentives is published alongside this Transition Plan.

Box 5
Could we generate power from the Severn Estuary?

The Severn Estuary has one of the largest tidal ranges in the world – a difference of about 14 metres between high and low tide. This could be harnessed to provide a long-term supply of electricity – up to 5% of the UK's supply – and could make an important contribution to delivering our renewable energy and climate change targets. But, as well as huge potential, a scheme in the Severn also has drawbacks including its costs and impacts on the natural environment.

The Severn Estuary

The Government is investigating these issues, and after public consultation in 2010, will decide whether to support a scheme in the Severn. A full planning process would follow a decision to go ahead. The potential for new innovative technologies is also being considered and the Government will consider the potential of new technology options before taking decisions on Severn tidal power.

Alongside this Transition Plan, the Government published a shortlist of schemes under consideration – including several options for barrages and impounded areas within the Estuary, and which innovative technologies will receive funding.

To deliver this support, the Government has launched the Office for Renewable Energy Deployment. The Office, part of the Department of Energy and Climate Change, will have the objective of significantly ramping up renewables deployment, by addressing delivery issues across a range of renewable energy technologies with a strong emphasis on facilitating investment and supporting the supply chain to maximise the economic opportunity presented by the UK renewables target.

Exploring untapped resources

The Government is undertaking work to explore the potential of untapped renewable sources, including the Severn Estuary – see box 5 for further details

Nuclear power

In January 2008, following consultation, the Government published its White Paper on Nuclear Power.[4] The White Paper sets out the Government's view that nuclear power is low carbon, affordable, dependable, safe and capable of increasing diversity of energy supply. The White Paper also explains the Government's belief that new nuclear power

4. http://www.berr.gov.uk/files/file43006.pdf

stations should have a role to play in this country's future energy mix alongside other low carbon sources; that it would be in the public interest to allow energy companies the option of investing in new nuclear power stations; and that the Government should take active steps to facilitate this.

Nuclear power is a proven technology, and it will be for energy companies to fund, develop and build new nuclear power stations in the UK, including meeting the full costs of decommissioning and their full share of waste management and disposal costs. However, the Government has a strategic role in removing unnecessary obstacles to the development of new nuclear power stations.

Since the White Paper on Nuclear Power, the Office for Nuclear Development, part of the Department of Energy and Climate Change, has made good progress in helping to facilitate nuclear new build:

- The Government is running a Strategic Siting Assessment to establish which sites in England and Wales are potentially suitable for the deployment of new nuclear power stations by the end of 2025. A list of potentially suitable sites will be included in a draft National Policy Statement for nuclear power, which will be published for consultation later this year (see below).
- The Government has legislated in the Energy Act 2008 to ensure that operators of new nuclear power stations will have secure financing arrangements in place to meet the full costs of decommissioning and their full share of waste management and disposal costs.
- The Government is running a process of Regulatory Justification to meet the requirement of European law under which Member States must make an assessment to ensure that the benefits of a new class or type of practice involving ionising radiation outweigh any detriments to health. The Government will be consulting on the Secretary of State's proposed decisions on this later in 2009.

Box 6
Safety and security of nuclear power

The UK has strict, independent, safety and environment protection regimes for nuclear power. Any new nuclear power station will be subject to safety licensing conditions and will have to comply with the safety and environmental conditions set by the regulators. Statutory obligations require that radiation exposures not only comply with dose limits but are as low as reasonably achievable. The security of civil nuclear material and sites is regulated by the Office of Civil Nuclear Security (OCNS). OCNS ensures that security measures are included in plans for the construction of any new nuclear power station from the outset.

Having reviewed the evidence and arguments, and based on the advice of the independent regulators, and the advances in the designs of power stations that might be proposed by energy companies, the Government believes that new nuclear power stations would pose very small risks to safety, security, health and proliferation. The Government also believes that the UK has an effective regulatory framework that ensures that these risks are minimised and sensibly managed by industry.

Nuclear is part of the future

- The Government is working to ensure that the nuclear regulators continue to have the resources and tools necessary to assess the safety, security and environmental impacts of new nuclear reactor designs through the Generic Design Assessment process by June 2011.
- In addition, the Government launched a consultation in June 2009 on proposals to enhance the transparency and accountability of nuclear regulation, creating a regulator with the autonomy and flexibility needed to meet challenges of the changing nuclear environment.

As set out in this plan, in order to decarbonise, our electricity supply will need to come from a mix including renewable sources, nuclear power and fossil fuels with carbon capture and storage. The draft National Policy Statement (NPS) for nuclear power, which the Government is publishing for public consultation and Parliamentary scrutiny later this year, will set out in more detail why the Government considers there is an early need for nuclear power as part of this mix.

The new Infrastructure Planning Commission, when set up, would use the NPS to help it make decisions about applications to develop new nuclear power stations on particular sites. This would allow for the earliest possible deployment of new nuclear power stations.

The action taken so far has resulted in real interest in new nuclear in the UK with energy companies announcing plans to build over 12 GW of new nuclear capacity. The Government is committed to enabling nuclear new build as soon as possible, and envisages the first new nuclear power stations operating from around 2018, but will look to accelerate timescales where possible.

The Office for Nuclear Development will continue to work with industry and others to meet new nuclear skills requirements and to develop a globally competitive UK supply chain, focusing on high value added activities to take advantage of the UK and worldwide nuclear new build programme.

The Government is also taking action to deal with our old nuclear facilities effectively and responsibly. The Government created the

Box 7
Managing the waste from new nuclear power stations

The Government is aware of the need to manage radioactive waste effectively and particularly the need to make progress towards a long-term disposal solution.

In June 2008 the Government published the White Paper on Managing Radioactive Waste Safely. This sets the framework for managing higher activity radioactive waste in the long term through geological disposal, coupled with safe and secure interim storage and ongoing research and development. It also invites communities to express an interest in opening up without commitment discussions with Government on the possibility of hosting a geological disposal facility at some point in the future.

As was made clear in the White Paper on Nuclear Power, our policy is that, before development consents for new nuclear power stations are granted, the Government will need to be satisfied that effective arrangements exist or will exist to manage and dispose of the waste they will produce. The Government currently expects to set out its view on this in the National Policy Statement for nuclear power.

Nuclear Decommissioning Authority (NDA) to ensure that civil public sector nuclear sites are decommissioned and cleaned up safely, securely, cost effectively and in ways that protect the environment and is funding the largest ever amount of expenditure on the UK civil nuclear clean-up programme.

In its first four years the NDA has made significant progress. For the first time there is a single body with responsibility for an ever improving understanding of the UK's nuclear liabilities and good progress has been made on decommissioning, with the focus being on tackling the highest hazards, particularly at Sellafield.

Clean fossil fuels through carbon capture and storage

Coal and gas will remain important to ensure our electricity supply is reliable and secure as we move towards a greater dependence on intermittent renewable sources like wind. However, coal power stations have higher carbon emissions for a unit of electricity than any other fuel and they can only remain part of our energy mix if they can be part of a low carbon future. There is the prospect of a solution to the challenge – carbon capture and storage (CCS) – which has the potential to reduce emissions from fossil fuel power stations by up to 90%.

Carbon capture and storage is the capture of CO_2 from large point sources such as, power stations and other industrial installations, transporting it and storing it underground. There are different ways of doing this – it is not one single technology.

One of the biggest challenges with this technology is that while each stage – capture, storage and transport – has been shown to work, CCS has never been tried at a commercial scale on a power station and never the complete process from start to finish. Action is therefore needed to demonstrate the technology at commercial scale, whilst ensuring the UK is prepared for its eventual deployment.

The UK is leading international efforts to develop CCS and, in 2007, we were one of the first countries to launch a commercial scale CCS demonstration project. In April 2009 the Government confirmed that subject to receiving suitable bids and being able to reach appropriate terms, it remains the Government's intention to proceed with the current competition to contract award. As with any long-term procurement, final funding approval for this will depend on decisions taken at the next Government Spending Review. The Government also announced public funding for the next stage of the competition and bids are now being selected to proceed with this stage. The Government confirmed all new combustion power stations over 300 MW in England and Wales would have to be designed and built "carbon capture ready" i.e. so that they could fit CCS. This will minimise the barriers to deployment of CCS once the technology has been proven.

But if we are to see CCS ready for commercial deployment from 2020, we need to go further. In April 2009, the Government also set out proposals for an ambitious new financial and regulatory framework for the development of carbon capture technology, followed by publication of a formal consultation in June, which closes on 9 September. The consultation sets out, and seeks views on, proposals to drive the development of clean coal by:

Providing financial support for up to four commercial-scale CCS demonstrations in Britain covering a range of CCS technologies including for example pre and post combustion capture.

- Requiring any new coal power station in England and Wales to demonstrate CCS on a defined part of its capacity: at least 300 MW net (around 400 MW gross).
- Requiring new coal power stations to retrofit CCS to their full capacity within five years of CCS being independently judged technically and economically proven. The Government will plan on the basis that CCS will be proven by 2020.
- Preparing for the possibility that CCS will not become proven as early as the Government expects.

The Government is focusing further actions on coal-fired power stations because the emissions per unit of electricity are substantially higher than from gas; these higher emissions mean that tackling coal first makes the most economic sense; and the projected increases in coal use globally create a greater sense of urgency to tackling emissions from coal. However, all gas plants (over 300 MW) will in future be built ready to fit the technology in order to facilitate its deployment once it has been proven.

Low carbon coal technologies represent a major future market for UK business estimated to be worth of the order of £2-4 billion to the UK by 2030, sustaining 30,000 - 60,000 jobs.[5]

The Government is considering how to encourage clusters of CCS infrastructure and expertise, in key areas, such as Yorkshire and Humber, the Thames Estuary, the Firth of Forth, Tyne/Tees and Merseyside, bringing major employment and regeneration benefits.

Working with other countries to develop CCS will be critical to developing and proving the technology because a large number and range of commercial scale demonstration projects will be needed globally to meet our climate change goals. The International Energy Agency's Technology Roadmap for CCS envisages 30 will be needed globally by 2020. Details about international action in this area are set out in Box 9.

5. AEA (2009) Future Value of Coal Carbon Abatement Technologies to UK Industry. www.decc.gov.uk"

Box 8
Office of Carbon Capture and Storage

CCS is a crucial part of the solution to climate change, in the UK and globally. It also offers an opportunity to create a new industry in capturing and storing carbon.

The UK is well placed to lead the way because the North Sea offers many potentially suitable sites for carbon storage. And it is possible to envisage a future in which there are geographic clusters of carbon capture, in areas like the Thames Gateway, Humberside, the Firth of Forth, Merseyside and Tyne/Teesside, based around pipelines transporting carbon dioxide to storage under the North and Irish Seas. The UK has the basics from which to build a CCS industry, from strong clean coal research and development expertise through a long tradition of fossil fuel power generation to the skills and experience of working with hydrocarbons on and offshore.

The UK is also home to the engineering and project management expertise needed to integrate the CCS chain of capture, transport and storage, and the financial and legal expertise to pull projects together. It is the job of Government to enable this future to become reality, helping to develop an integrated capability to take forward these very major projects at home and overseas.

The measures proposed in the recent consultation document, published on 17 June, to support up to four demonstration plants are a start. But they need to be the start of a road to deploying CCS technology at scale, so creating the new industry of the future. So the Department of Energy and Climate Change now intend to establish an Office of Carbon Capture and Storage ("OCCS") within the Department to make this happen, in the same way that it has created an Office for Nuclear Development and an Office Renewable Energy Deployment.

The first step will be to consult with stakeholders on the objectives, functions and tasks of the OCCS. The Department will aim to announce further details of the new Office in the autumn.

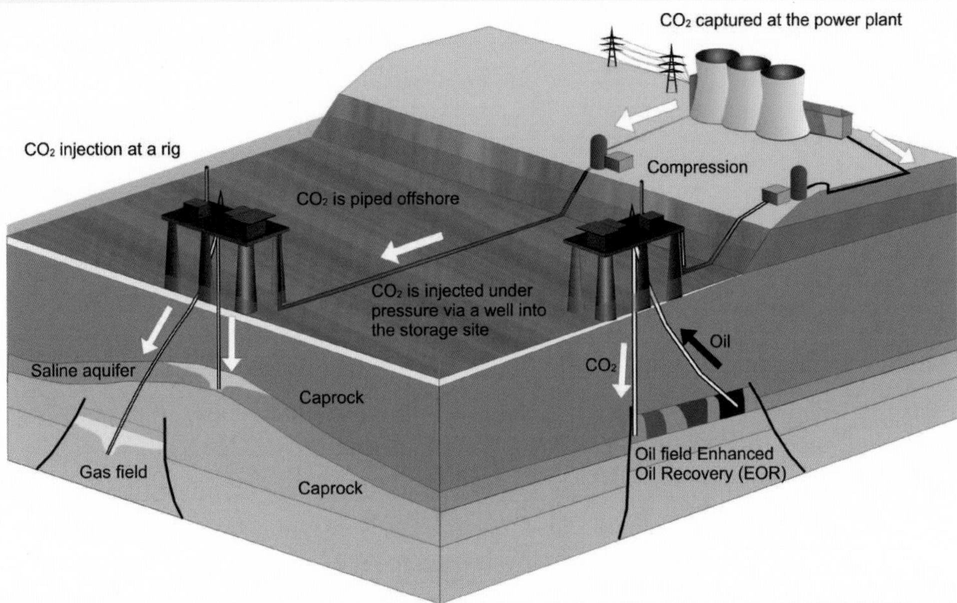

Some carbon capture and storage techniques involve capturing the carbon dioxide emissions from fossil fuels, and storing them underground. British Geological Survey (2008)

Box 9
The UK is working internationally to develop carbon capture and storage

- The UK is working with EU partners to develop CCS and in December 2008 was instrumental in reaching agreement that allowances from the EU Emissions Trading System should be used to support the EU's ambition to have 12 demonstration projects by 2015. Further support for CCS projects in seven EU member states was agreed in April 2009.
- In July 2008, the G8 leaders announced that "we strongly support the launching of 20 large scale CCS demonstration projects globally by 2010, taking into account various national circumstances, with a view to beginning broad deployment of CCS by 2020". The UK will be seeking concrete progress this year on this commitment.
- The UK has also played a key role in the development of the EU-China Near Zero Emissions Coal Initiative to demonstrate commercial scale CCS in China.
- To facilitate international progress on CCS, the UK is co-hosting with Norway the Carbon Sequestration Leadership Forum Ministerial meeting on 12-13 October 2009 in London, where Ministers and CEOs from around the world will convene to agree steps to bring forward the commercialisation of CCS.

Taking action to create a bigger, smarter grid

Whatever combination of technologies exists from now to 2050, we will need a smarter, more flexible grid that will be able to manage electricity generated from new technologies and respond to changes in energy demand. In the shorter term the Government needs to ensure that the grid is able to respond to the pressures for connecting the new power stations under construction in a timely way.

More investment in the existing grid

The UK network needs expansion to connect the new sources of generation the Government needs to meet our carbon and energy goals in 2020. Ofgem has approved £4-5 billion of refurbishment and expansion plans to ensure the grid is capable of supplying current and future electricity to our homes, businesses and industry in the next few years. The Government is developing a Grid National Policy Statement and will publish it for consultation later this year.

Looking towards 2020, the DECC-Ofgem co-chaired industry group: Electricity Networks Strategy Group (ENSG) has produced a vision for network investments the Government needs to meet 2020 renewables and energy targets. The ENSG found that an additional £4.7 billion was needed to develop the onshore grid to support connection of up to 35 GW of renewable generation coming forward by 2020 and accommodate increased flows of electricity across the network. This work presented a consensus view of the strategic investments that are required under different scenarios and an agreed plan to deliver these investments ahead of requests by generators for a grid connection. Ofgem has approved up to £43 million of pre-construction work on projects identified in the ENSG vision and by summer 2010 will finalise new incentives for network companies to build the new lines necessary to meet 2020 targets.

The Government also needs to consider now how to plan for network investment beyond 2020. As set out in Chapter 8, Government is considering what further actions are necessary in shaping the long term path to meeting our energy and carbon policy objectives. A key element of the future energy system is the network infrastructure required to support and match the demand for and supply of electricity.

This includes how the Government should work with the regulator and in consultation with the industry to develop a long term vision for the network, including the investments that would help contribute to the Government's policy objectives. Part of the Government's considerations will include how best to underpin this vision and how all parties can report progress on their contribution to the delivery of the vision. This work will address questions about the scale of the investment required; the prioritisation of investment; and the investment needed to produce a smarter grid capability, consistent with the interests of consumers today and in the future.

Setting the right incentives framework for network companies will also help deliver networks which are fit for purpose. Through the next electricity distribution price control to be set from April 2010, Ofgem is looking to expand funding for innovation and to get the networks ready for the challenges of the low carbon future such as more local generation. In addition, Ofgem's RPI@20 project will report in summer 2010. This comprehensive review covers all networks to make sure the regulatory framework is capable of delivering the networks needed for a low carbon economy.

Quicker and fairer connection to the grid

Timely and effective reforms of the grid access regime are essential to speed up the connection of new generation and to ensure security of supply. A new grid access regime is required which is consistent with the Government's policy goals for meeting the renewable energy target in 2020, as well as its broader energy policy objectives. The Government therefore took powers in the Energy Act 2008 to implement new grid access rules if the industry grid access reform process did not deliver as quickly as needed.

Having considered industry submitted proposals for enduring grid access reform, Government has decided to use the Energy Act 2008 powers to implement

Offshore wind farms will be part of the future electricity supply

enduring reform of the transmission access arrangements. The Government has considered the industry proposals and the advice of Ofgem on the best way forward. The Government's key considerations have been the extent to which the proposals add certainty for developers, speed up connections, and ensure the efficient operation of existing plant. The risks to current and future consumers of grid access reform have been considered in relation to the impact on the required generation investment, and other Government policy goals, and the costs of the possible options.

The Government has decided to use the powers taken in the Energy Act 2008 to introduce the necessary reforms. To make progress in the short term, the Government has also taken urgent steps - with Ofgem, National Grid and industry - to ensure that ready to go projects can connect at the earliest possible date. This has successfully resulted in around 1 GW of renewable projects receiving earlier connection offers – enough to power 600,000 homes.

Investment in a new offshore grid

The Government and Ofgem have developed a new regulatory framework for offshore electricity transmission to provide clear, cost effective and co-ordinated delivery of £15 billion worth of grid connections needed for

Box 10
Key elements of a UK smart grid

- Improved information for electricity consumers, notably through smart meters, to allow them to manage their energy use (and hence energy bills) more effectively.
- Facilitating demand management, providing data to technologies in homes and buildings that can regulate electricity use (e.g. encouraging electric cars to recharge when there is "surplus electricity" available on the system).
- Enabling individuals and businesses to sell electricity into the network as well as buying from it, through microgeneration and on-site technologies.
- Enhanced monitoring and information flows for network operators, allowing them to make more efficient decisions about where energy flows across the network on a real time basis. This is likely to be particularly important with increasing levels of intermittent renewable generation on the system. A greater use of energy storage would also increase the need for smarter information flows for network operators on energy storage supply and timing of its use.
- Use of a range of technologies including advanced communications and information management systems, intelligent metering, demand side management, and storage. Many of the technologies to enable such capability are already available, but have not yet been integrated together in large scale demonstrations and the actual mix that is deployed will depend on their feasibility.
- More optimal usage of the whole network in meeting demand, which could limit the need for more reinforcement of the grid.

radical offshore wind growth. This regime went active in June 2009.

A smarter grid for the future

By 2030 we will see a step change in intermittent generation delivered through both large and small scale renewable plants, more price responsive consumers enabled by the roll out of smart meters and possible increases in demand for electricity from structural changes in power use, such as a move to use of electric vehicles.

In the long term the electricity grid will continue to need to develop in order to connect and integrate new technologies and enable even more active management of fluctuations in supply and demand than we experience today. To do this network operators could benefit from having more real time information on energy use and supply and network conditions and capability for more automated response to changes on the network.

The Government has therefore started considering what a UK smart grid might look like in practice; possible key elements are set out in box 10.

The UK already has a grid capable of coping with fluctuations and balancing supply and demand. The key issue is developing this capability even further to cope with the evolving energy system to 2050 so that the grid is fit for purpose in the future. The Government is already working on key areas of smart networks to ensure it understands the benefits of smart grid technologies for meeting our goals and to encourage deployment in the coming years. Its programme currently consists of:

- Smart meters: A programme to roll out smart meters to every home by end 2020 – an £8bn private sector investment. This is one of the building blocks for creating a smart grid (see Chapter 4).
- Deploying new technologies: Encouraging Distribution Network Operators through regulatory incentives to trial new 'smarter' technologies on their networks. Ofgem are proposing to significantly increase the amount of funding available for this.
- Developing new technologies: Providing direct funding for innovation through the Energy Technologies Institute (ETI) which aims to invest up to £1bn over the next 10 years in low carbon energy technologies, including networks. The Government is engaging with the recently established ETI networks panel that is scoping the objectives they set for a call for projects.
- Funding research: Providing direct funding, through the Research Councils, of over £30m for collaborative research in networks involving academia and industry. Providing complementary funding of £6m to supplement other funding for network innovation such as Ofgem's Innovation Funding Incentive amongst other sources. Government funding for smart grids will be used to support early stage development of trials of key technologies consistent with a vision for a smart grid in the UK to be published later this year.

The costs and benefits of a smart grid will ultimately depend on the combinations of technologies that are brought together - some are well understood, some at an early stage of development, others do not yet exist.

The ENSG 'smart grid working group' are taking forward a study on smart grids in a UK context. The study will consider the costs, benefits and issues to be addressed in developing a smart grid for the UK electricity system, including technology readiness, how such a system might develop, and the drivers and barriers at each stage.

Following completion of this work the Government will publish:

- A high level "vision" setting out what a UK smart grid might look like and why the Government would want to develop it later this year.
- In the light of work on the 2050 energy roadmap next year, a 'route map' for delivery of this vision.

Using electricity more efficiently

An effective way of reducing energy demand is through the introduction of minimum energy efficiency standards for new products on sale. European Union countries agree the rules on minimum energy efficiency standards for products such as white goods and televisions; and the information available to consumers, such as the 'A-G' energy label rating. These minimum standards mean that the products which waste the most energy are taken off the shelves.

The Government is pushing to achieve more, and extend the ambition set out in the 2007 Energy White Paper, with the aim of doubling the emissions savings expected by the measures agreed so far, by 2020. To achieve this, Government will continue to work hard in Europe in the next two years to ensure that new standards are brought forward.[6]

The minimum standards agreed so far in Europe are expected to save £900 million per year from people's energy bills and help save around 7 MtCO2 per annum by 2020. In the UK, the Government has also introduced a voluntary initiative to bring forward the phase-out of energy-wasting incandescent light bulbs by 2011, ahead of EU legislation.

Keeping our electricity supplies safe and secure

To date, our electricity system has given us an extremely reliable supply, with the average consumer in the UK spending less than an hour and a half without power in a year. It is critical that the UK continues to have secure and reliable supplies of electricity as we make the transition to a low carbon economy. However, the scale and pace of the transformation that we expect to see in the electricity sector means that we need to be alert to new challenges to our electricity security of supply as we make this transition.

The Government's Updated Energy Projections suggest that demand for centrally generated electricity will be lower over the next decade compared to today given the combined impact of our energy efficiency policies, the increase in small scale generation and the economic downturn. However, it is possible that beyond this electricity demand could be significantly higher, for example we may use electricity for transport or to heat our homes.

Ensuring security of supply during our transformation to lower carbon electricity is a particular challenge because of the lead times for building new power stations and the requirement for significant capital investment.

To deliver secure supplies of electricity requires:

- a supportive climate for timely investment in a diverse mix of low carbon technologies; and
- a market and regulatory framework that can adapt to the different characteristics of low carbon electricity generation technologies.

6. The analysis in this plan does not take account of this potential additional abatement which the Government will consult on later this year.

Outlook to 2020

Around sixteen power stations representing approximately 25% (18 GW) of our electricity generating capacity are scheduled to close by 2018. Significant new investment is needed to replace them and to ensure a healthy margin of spare electricity generation capacity, and latest figures show the market is responding to this need with over 20 GW of investment under construction or with planning consent. These investments comprise the following:

- The construction of 2 GW of generating capacity that has recently been completed and will be commissioned this year.
- 8 GW of new generating capacity that is currently under construction
- An additional 10.5 GW that has both planning consent and agreement to connect to the grid
- A further 7.5 GW that has applied for planning consent in England and Wales

Chart 5

The UK is likely to have sufficient generating capacity in the mid-teens despite power station closures

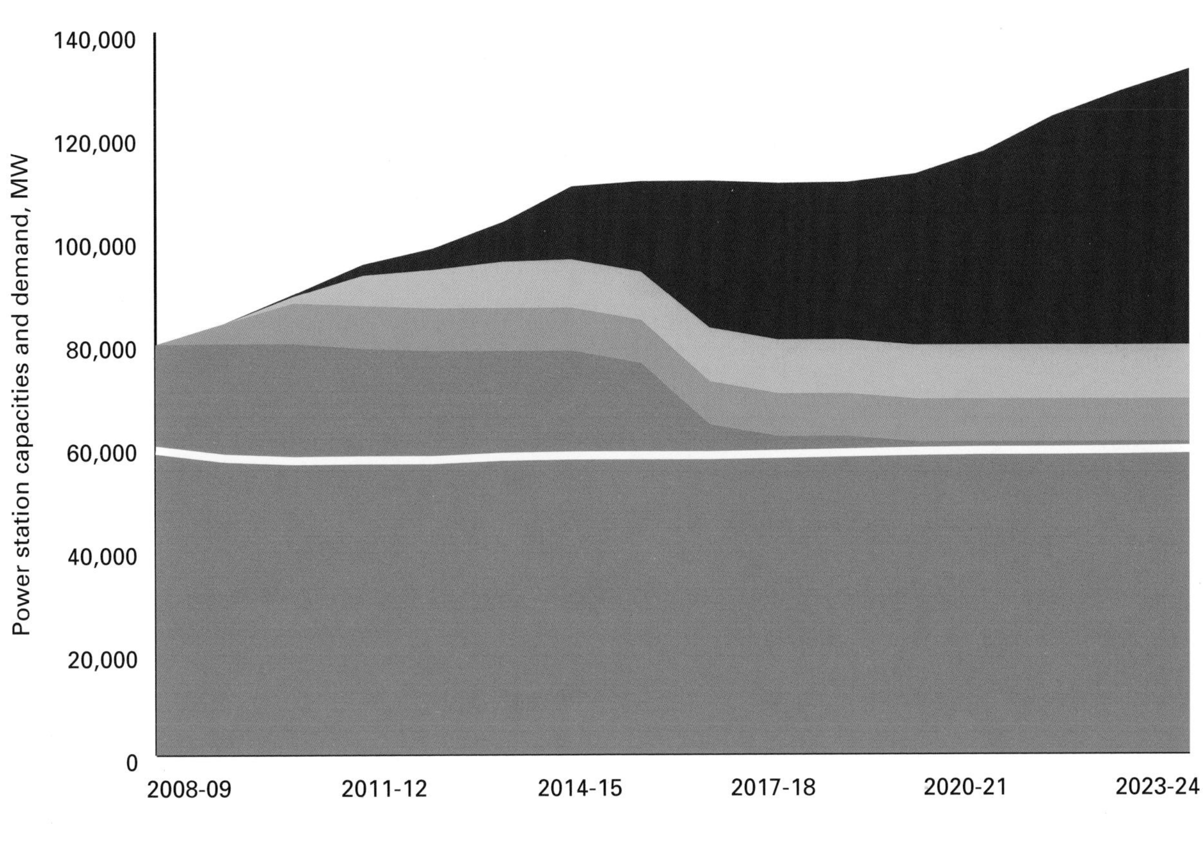

* Transmission contracted

Source: National Grid Seven Year Statement extrapolated to 2023; Department of Energy and Climate Change

Chart 5 shows the overall impact of these changes on electricity supply, compared to the National Grid's projection of demand.

Beyond the middle of the next decade, further closures will be driven by the EU's Industrial Emissions (Integrated Pollution Prevention and Control) Directive, which will replace the regulatory framework established by the EU's Large Combustion Plants Directive and sets stricter limits on the emissions of sulphur and nitrogen oxide. The Government believes that the position reached at the Environment Council in June 2009 would enable us to manage the risks that these further closures could pose to security of electricity supplies while allowing a smooth transition to a system of low carbon generation. The Directive now needs to be agreed by the European Parliament before it becomes law.

Renewable electricity can be generated from ocean waves

A diverse mix of low carbon technologies

A diverse mix of low carbon technologies helps deliver energy security by reducing the risk of problems that may arise with one type of technology or fuel. The challenge is to ensure that these technologies, each of which has different characteristics, can operate together to provide reliable secure supplies of electricity.

A significant share of our future electricity supply is likely to be generated from renewable sources (e.g. wind and tidal). Wind generation, which is likely to provide most of our renewable generation in 2020, is both intermittent and unpredictable; so we will need power stations whose output can be flexed at short notice to ensure that demand can be met during periods when the wind is not blowing.

As the output of nuclear power stations is relatively inflexible, in the medium term this flexibility is likely to be provided by fossil fuel fired power stations. Fossil fuel fired power stations are able to quickly change the amount of electricity they generate and this flexibility can be used to ensure a reliable and predictable total supply of electricity to meet demand.

Longer term there are a number of options for making our electricity system more flexible, while at the same time ensuring a shift to low carbon supplies, including: coal and gas fired power stations with carbon capture and storage; greater interconnection; more storage capacity; and greater flexibility in electricity demand, which could be provided by active management of new sources of electricity demand such as electric cars.

Box 11
How renewable electricity affects wholesale electricity prices

Increased levels of renewable electricity generation are likely to change the shape of electricity prices. Independent analysis[7] suggests that the volatility of spot prices is likely to increase as a result of wind generation. The analysis shows that wholesale prices may fluctuate from levels which are sometimes negative (due to wind generation bidding at its opportunity cost of -1 ROC) to above £1,000/MWh. The probability of lower and negative prices is likely to increase as well as the probability of higher peak prices so that on average wholesale electricity prices and hence retail prices will not necessarily increase. The analysis also suggests that this volatility could dramatically increase by 2030. By comparison, spot prices reached a high of around £500/MWh last year (the Analytical Annex provides additional details).

A supportive climate for investment

To ensure secure supplies while reducing power sector emissions we will need a market and regulatory framework that supports investment in increasingly low carbon generation with sufficient flexibility. Factors affecting power station investment decisions include expected fossil fuel, carbon and electricity prices. Fossil fuel prices are influenced by international markets, while the Government sets the framework which delivers a price for carbon through the EU Emissions Trading Scheme. The growth of renewable electricity is expected to make future electricity prices more volatile – see box 1.

Higher wholesale prices for electricity at times of peak demand will be important in providing sufficient returns for investors in the flexible power stations. These stations might be expected to operate for a limited number of hours per year when renewable electricity generation is not available. So these investors will need to be confident that prices will reach sufficiently high levels on a sufficient number of occasions to allow them to recover their costs.

Our analysis suggests that we will see sufficient investment in flexible back up generation to ensure secure supplies so long as investors expect prices to accurately reflect the value of providing this flexible capacity.

National Grid's recent consultation[8] sets out their view that a number of changes to the technical operation of the supply and demand balancing arrangements might be necessary in the period to 2020 in part to manage the intermittency of renewables supplies. There would need to be greater reserve capacity, but they did not consider that fundamental change was required.

The Government believes that the risks to electricity security of supply are manageable to 2020 given its assessment of:

- future power station closures, the investments coming forward and future electricity demand, and
- the implications for electricity security of supply from the increase in intermittent wind generation delivered by the Government's Renewable Energy Strategy (2009).

However, the Government recognises that the scale and pace of change needed will test our market during the transition to a low carbon economy and some issues may need to be considered further. For example,

7. Impact of Intermittency: How wind variability could change the shape of the British and Irish electricity market. Pöyry. July 2009.
8. 'Operating the Electricity Transmission Networks in 2020'

the amount of renewable electricity on the system could potentially become a problem after 2020 due to the closure of the older gas and coal power stations which provide valuable flexibility to the electricity system.

There may also be impacts on other forms of low carbon generation (including nuclear and CCS); if investors expect future electricity prices to be more volatile this could increase their uncertainty over the expected returns from investing in these technologies.

The Government will need to be vigilant to ensure that the market framework works effectively. As levels of renewable electricity generation increase, the Government will work closely with Ofgem and in consultation with the industry and other stakeholders to consider whether further steps might be necessary to address these. The Government will seek stakeholders' views on our assessment of these electricity market issues in a call evidence on electricity this summer including the issue of how greater demand side management can be exploited. The Government will consider responses as part of longer term planning under the roadmap work – see chapter 8.

Security of supply of fossil fuels

In addition to these actions, the UK needs to maintain security of supplies of fossil fuels, which will remain an essential input to our electricity supplies for many years to come. For example, average annual coal burn for generation is around 50 million tonnes. Around a third of this is supplied by the UK coal industry, which produced 17.4 million tonnes in 2008. The Government is acting to maintain security of supplies both within the UK and internationally. Chapter 4 sets out the Government's view on the outlook for security of gas supplies, including Government action internationally to build good relationships with other countries to help ensure our supplies are secure.

Chapter 4
Transforming our homes and communities

Summary

Over three quarters of the energy we use in our homes is for heating our rooms and water, most of which comes from gas-fired boilers. This accounts for 13% of the UK's greenhouse gas emissions, but by 2050 emissions from homes need to be almost zero.

The Transition Plan, along with wider policies, will cut emissions from homes by 29% on 2008 levels. It will also improve the support to the most vulnerable, and secure the UK's gas supplies as our indigenous reserves in the UK Continental Shelf decline. To do all this, the Plan will help us will reduce the amount of energy we use, and produce more of our heat and electricity in a low carbon way, using technologies such as solar power and heat pumps.

Highlights of the Transition Plan include immediate help for households to make energy savings, helping to meet the commitment to insulate six million homes:

- Increasing the obligation on energy suppliers to help households reduce emissions and save energy – the Carbon Emissions Reduction Target – by 20% between April 2008 and March 2011, so that about £3.2 billion will be invested; and confirming that an obligation on energy suppliers will be maintained further into the future, up to the end of 2012, meaning around 1.5 million additional households will benefit from significant energy saving measures on top of the six million commitment

For delivering energy savings in the longer term, the Plan will:

- Support households to take action by:
  - Installing smart meters in every home by the end of 2020, which will enable people to understand their energy use, maximise opportunities for energy saving, and offer better services from energy companies.
  - Encouraging the provision of smart displays now for existing meters benefitting some two to three million households, and launching a new personal carbon challenge with rewards and incentives for saving energy.

Chart 1
Our homes and communities will contribute 13% of the additional savings in 2018-22

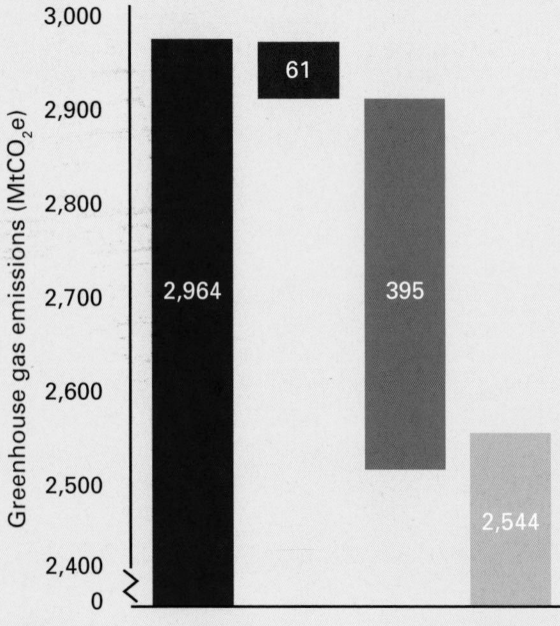

Emissions before policies
Savings from homes & communities
Savings from other sectors
Carbon Budget 2018-2022

Note: Reductions due to policies introduced prior to the Energy White Paper 2007 are not shown.

Note: Savings also include interaction effects. Refer to Table A1 for a full breakdown of carbon savings in the third budget period

Source: Department of Energy and Climate Change analysis (2009)

  - Developing more proactive services from the Energy Saving Trust to provide households with information and advice at the right times.
- Help people meet the costs of transformation by:
  - Piloting a move from upfront payment to "pay as you save" models of long term financing for energy saving, so it will be more affordable to make the changes needed to make the whole house low carbon
  - Introducing clean energy cash-back schemes so that people and businesses will be paid if they use low carbon sources to generate heat or electricity, meaning a household with a well-sited solar panels could receive over £800 plus bill savings of around £140 a year.
- Coordinate the support available by:
  - Introducing a community-based approach to delivering treatments to homes in low-income areas, through the Community Energy Saving Programme.
  - Considering the delivery mechanisms that will best deliver significant whole house energy saving treatments in the longer term, setting out the strategy later this year.
- Raise standards in every home by:
  - Consulting on requiring Energy Performance Certificate ratings for rented properties to be put on property advertisements; and consulting on extending access to EPC information to help target energy efficiency offers and support.
  - Requiring new-build homes to be built to high environmental standards, reaching 'zero carbon homes' from 2016.
- Help communities to act together by:
  - Announcing £10m for 'Green villages, towns and cities', a challenge for communities to be at the forefront of pioneering green initiatives, with 15 or so 'test hub' areas.
  - Developing an online 'How to' guide for anyone looking to install renewable and low-carbon energy generating technologies at community scale.
  - Exploring how to unlock greater action by local authorities in identifying the best potential for low carbon community-scale solutions in their areas.
- By 2020 the impact of these measures will be to add, on average, an additional 6% to today's household bills. Including all our previously announced climate policies will increase this figure to 8%. This Transition Plan will help the most vulnerable with their energy bills by:
  - Creating mandated social price support at the earliest opportunity with increased resources compared to the current voluntary system. The Government is minded to focus new resources particularly on older pensioner households on the lowest incomes.
  - Increasing the level of Warm Front grants so most eligible applicants get their energy saving measures without having to contribute payment themselves.
  - Working to ensure that fuel poor households can benefit from new schemes, such as the Renewable Heat Incentive, to help reduce bills.
- Securing the gas the UK needs during the transition to low carbon, by maximising economic production of UK reserves, improving our capacity to import and to store gas, and having in place the strategic partnerships to source gas imports. The Government will shortly be issuing a commentary on gas security of supply.

The scale of the task

Households directly contribute to climate change through the energy we use to heat our homes and our water; and indirectly through other resource use including electricity used to power appliances (see chapter 3) and through the energy to treat and supply our water. Our use of gas to heat our homes also exposes us to energy security issues as indigenous gas supplies decline. The cost of meeting our heating needs can be difficult for some households to afford.

Chart 2

Over three-quarters of the energy we use in our homes is for heating

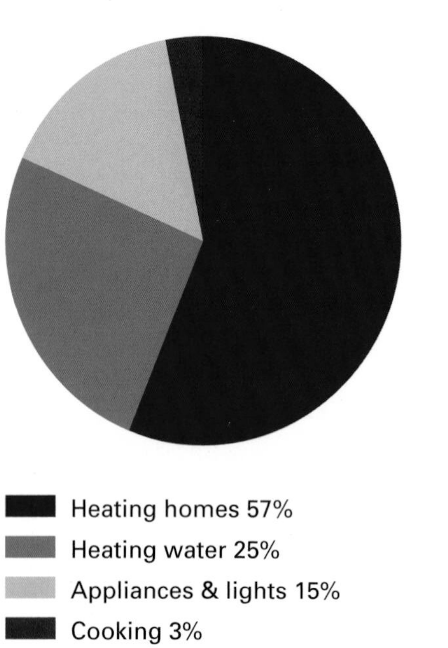

Heating homes 57%
Heating water 25%
Appliances & lights 15%
Cooking 3%

Source: Department of Energy and Climate Change, Energy Trends (September 2008)

Chart 3

Heating our homes and hot water is responsible for 13% of the UK's total greenhouse gases emissions[1]

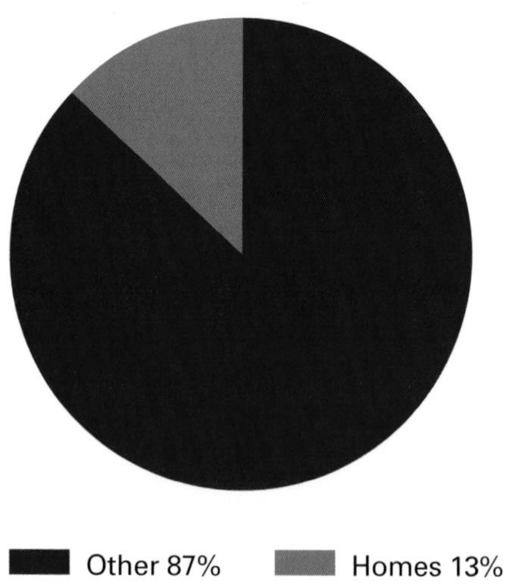

Other 87% Homes 13%

Source: National Communication source (2007)

The energy that households use to heat their homes and water accounts for over 80% of total household energy use (see chart 2), and the emissions from this account for 13% of the UK's total emissions (chart 3).

By 2050, the UK will need to have cut its greenhouse gas emissions by four-fifths overall. Looking across the options available, that means we will need to radically reduce demand for energy and decarbonise the energy we use in our homes almost totally by 2050. Our homes need to become much more energy efficient and we need to produce more of our heat and electricity from low carbon sources, such as ground and air-source heat pumps and solar power.

1. The emission estimates in this chapter refer to greenhouse gas emissions from combustion of (gas, oil, and coal) in residential properties. These fuels meet most of our heating needs for our homes and hot water and a proportion our energy use for cooking. Around 7% of domestic heating needs are met with electricity, the emissions from which are covered in the power chapter

The first Code for Sustainable Homes Level 6, zero carbon homes, in Upton Northamptonshire built for Metropolitan Housing Partnership

Much of our existing housing stock are period properties that tend to be draughty, but the new homes we are building are more efficient

Chart 4

The domestic energy efficiency package will deliver over two-thirds of emission savings from homes

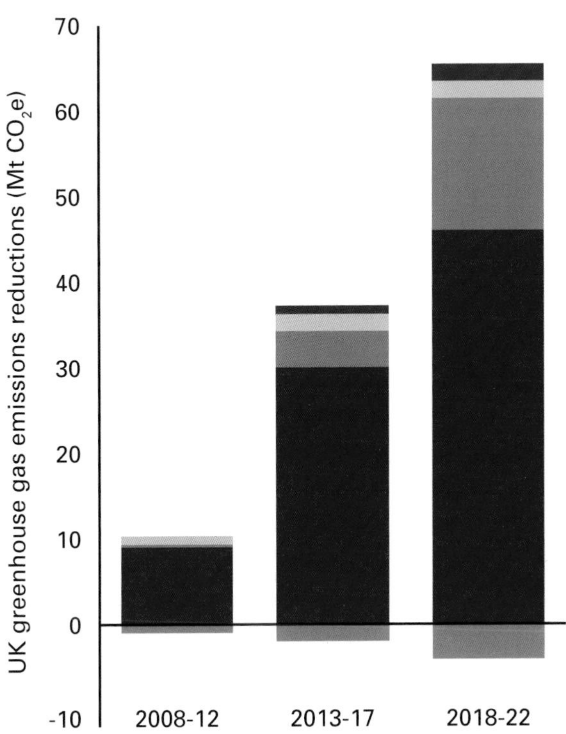

- Domestic energy efficiency (Carbon Emissions Reduction Target extension, future Supplier Obligation, Heat and Energy Saving Strategy and supporting measures)
- Clean energy cash-back and Renewable Heat Incentive supporting measures
- Smart metering and better billing
- Zero carbon homes
- Community Energy Savings Programme
- Additional product policy (see below)

Note: Reductions due to policies introduced prior to the Energy White Paper 2007 are not shown.

Note: Product policy savings are negative because of the Heat Replacement Effect. More energy efficient products create less ambient heat which needs replacing via alternative fuel sources. Overall, products policy provides a significant net benefit, due to savings in emissions in the traded sector and their associated benefits.

Source: Department of Energy and Climate Change

All households will need to play a part in this transformation. But we recognise that there are a number of obstacles to overcome in making the transition:

- Our current system works well, and because we are used to the way our boilers and appliances work, we have little appetite for change. Many of us are not aware of the options available, and we tend not to want to spend our time researching them to find out.
- Many of the changes we need to make to our buildings can involve disruption and cost. For example solid wall insulation can mean the need to redecorate rooms and can sometimes mean losing floor space. The cost of improving the energy performance of buildings has to be balanced against other priorities, even if we get the money back over time through savings on our energy bills. People living in rented accommodation face a further obstacle, if they are not able to make changes to the property.
- Modernising our buildings will not in itself be enough – we will also need to change our habits. Energy and resource use is not very visible and most of us have little idea of how much energy we use for different things. It is not always easy for people to see how small individual actions can make a difference. Sometimes people can be more effective by working together as a community.
- The Government wants to help people afford the energy they need to keep warm, but it can be difficult to reach those people experiencing fuel poverty who most need help.[2]
- During the transition to low carbon, gas will continue to be a significant source of heating. The UK's gas supplies are declining and we will need to make better use of it and good use of imports.

The plan to 2020

The policies set out in this Transition Plan will ensure that by 2020, the emissions from heating our homes will fall by 29% on 2008 levels; and that the most vulnerable people are protected from fuel poverty, and gas supplies for heating homes and other uses, are secure. The measures in this Plan will contribute 13% of the emission savings in the third budget period (see chart 4). Because of the challenges in this task, the Government must help to enable households to transform home energy efficiency and promote low carbon energy.

Earlier this year, the Government had a detailed public dialogue on the plans to transform home energy efficiency and provide low carbon energy. The majority of the people involved were overwhelmingly positive about improving their homes and using low carbon technologies, in principle. Participants said they look to the Government to take the lead, and are ready for some bold steps, but want information and advice, and help meeting financial costs.[3]

By 2020 we will cut emissions from homes to 29% below 2008 levels

2 The Fuel Poverty Strategy sets out a definition of a fuel poor household as being one that needs to spend more than 10% of its income on fuel to maintain a satisfactory heating regime (usually 21 degrees for the main living area, and 18 degrees for other occupied rooms).

3. For more information on this dialogue, visit our Big Energy Shift website, www.bigenergyshift.org.uk

A lot has already been done – for example almost two-thirds of cavity walls are now filled in the UK and 35% of lofts are insulated to at least 150mm. Warm Front fits or repairs a central heating system every minute of every working day in vulnerable households across England. But it is clear there is a long way to go to make progress to 2020. This Transition Plan sets out below action being taken:

- **Immediate help for households to make energy savings** through subsidised energy saving and insulation measures
- **Delivering energy savings in the longer term**
  - **Information to encourage action** with proactive advice services and installation of smart meters
  - **Helping people meet the costs of transformation** piloting new finance arrangements for home energy improvements and payment for low carbon energy
  - **Coordinating the support available** to explore whether new approaches to delivery will be better tailored to the scale of challenge
  - **Raising standards in every home**
  - **Helping communities to act together** through collective action and local authority involvement
- **Protecting consumers and helping the most vulnerable** with their energy bills to tackle fuel poverty.
- **Securing our gas supplies** in the transition to a low carbon future.

Putting the plan into practice

Immediate help for households to make energy savings

In September 2008, the Prime Minister announced the ambitious £1 billion Home Energy Saving Programme, to help families permanently cut their energy bills and to increase the rate of energy saving and carbon reduction, committing to insulate six million homes by the end of 2011. Reducing demand for energy will also reduce our need to import gas.

In February 2009, the Government set out its proposals to significantly increase the amount of help householders can receive to take up energy efficiency measures over the next few years, by requiring energy supply companies to offer more support to consumers:

- **The Government has increased by 20% the amount of support energy supply companies will make available through the Carbon Emissions Reduction Target (CERT).** The scheme has already helped over six million households with significant energy saving measures since 2002. The suppliers will now have to meet an increased level of emissions reductions,[4] meaning that an extra £0.6 billion is likely to be spent, bringing total support between 2008 and 2011 to an estimated £3.2 billion. Of this, it is estimated that around £1.9 billion will be directed at vulnerable households.

4. Increasing the obligation on suppliers from lifetime carbon savings of 154 million tonnes carbon dioxide saved to 185 million tonnes.

The Government is committed to helping six million households insulate their homes by the end of 2011

- To give consumers more opportunity to benefit from this support, **the Government now confirms following consultation that it will maintain an obligation on energy suppliers further into the immediate future, extending the end date of the CERT from April 2011 to the end of 2012.**[5] The suppliers' target for this period will be at least as ambitious as the current scheme on a pro rata basis, meaning around 1.5 million additional households could benefit on top of the six million commitment. The Government will propose that during this additional period there be a particular focus on insulation measures; and any early action which exceeds existing obligations under CERT will be proposed to be eligible towards an obligation for the extended period. The Government will also explore how best to provide help to those most vulnerable, including those living in fuel poverty. *The Government aims to publish a consultation on the details of this by the end of 2009.* This will provide industry partners with greater certainty so that they can make the necessary business planning and investment decisions to deliver these measures.

Delivering energy savings in the longer term

Over the longer term, as the potential for first steps like loft insulation and filling cavity walls has been taken up, we will need more substantial changes to existing homes and new homes, such as solid wall insulation and new low carbon sources of heat supply and electricity. The Government intends for all lofts and cavity walls to be insulated where practicable and where households want it by 2015.

In February 2009, the Government announced the "Great British Refurb" and launched its consultation on long-term plans for heat and energy saving.[6] A summary of responses to the consultation has been published; and later in 2009 the Government will set out its strategy. The consultation set out the Government's ambition that by 2030 all homes will have undergone a 'whole house' package including all cost-effective energy saving measures, plus renewable and low-carbon heat and electricity measures as appropriate.

Supplying the materials and installing the insulation and other energy saving measures into millions of people's homes will be a big task and an employment and economic opportunity for the green building sector and its supply chain over the medium term. Developing and deploying new technologies for low carbon energy and energy efficiency at household and community level will also present significant opportunities for UK industry.

5. Department for Energy and Climate Change (2009) Amendments to the Carbon. Emissions Reduction Target, http://www.decc.gov.uk/en/content/cms/consultations/open/cert/cert.aspx
6. *The Heat and Energy Saving Strategy Consultation* http://hes.decc.gov.uk/consultation

Figure 1:
The 'whole house' approach

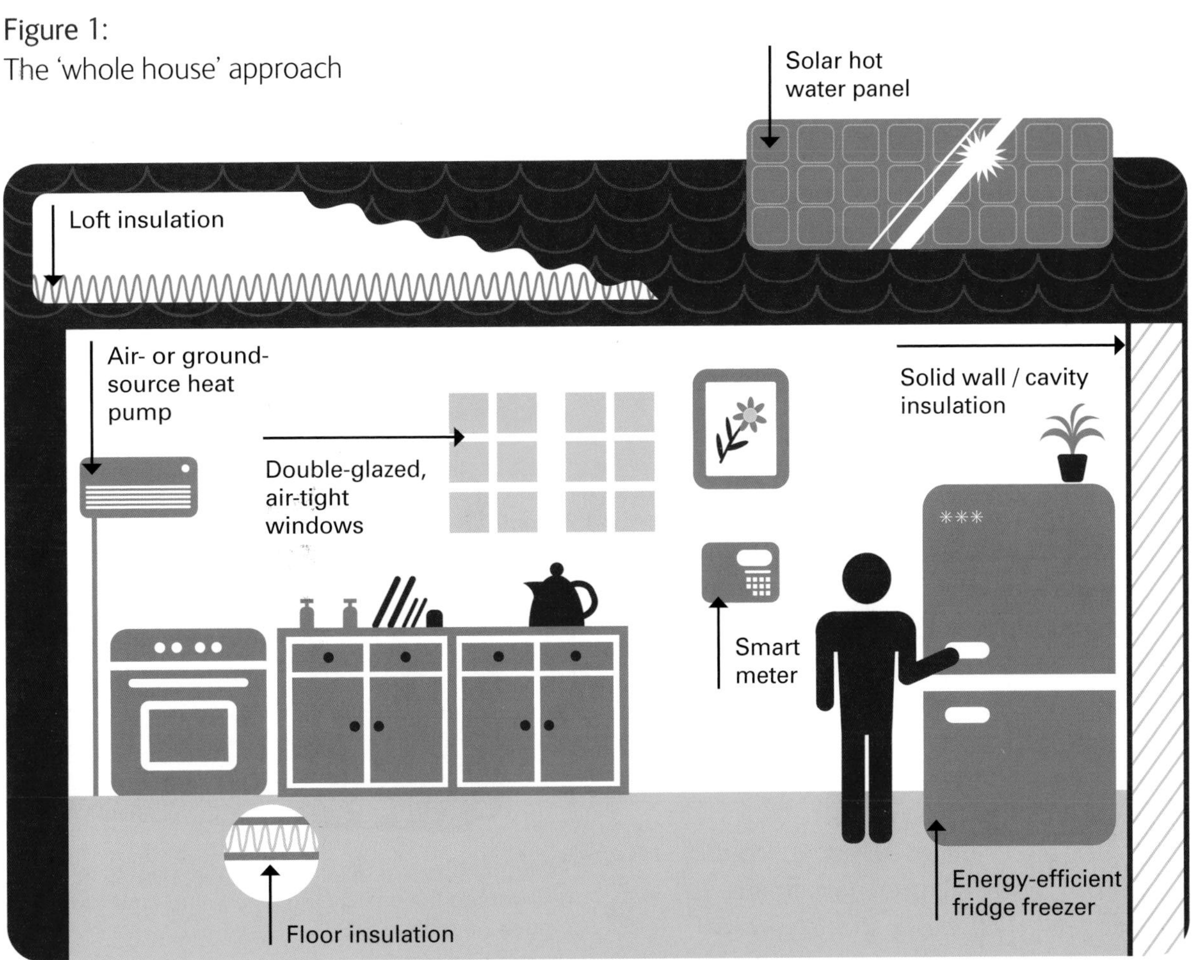

Box 1
The 'whole house' approach

A 'whole house' approach means considering a household's energy needs and carbon dioxide impacts as a whole, and establishing a comprehensive package of measures to address them.

The aim would be to include all the measures available that are suitable for a property and which could pay back through energy bill savings over their lifetime. This should result in a coordinated package, which will also include renewable energy measures where appropriate to the property.

A key benefit of the 'whole house' approach is that it ensures that the needs of the property are assessed as a whole, that they happen in the right order, and that disruption is minimised.

Providing information to encourage action

To encourage action, the Government needs to make sure that people have the right information available to prompt interest in energy savings, and impartial and accurate advice at the right time to enable householders to undertake the changes to their home that will help cut carbon, energy bills and our need for gas.

Smart meters are key to revolutionising customer service and maximising energy-saving. **The Government has committed to mandating smart meters and has set out an indicative timetable for getting smart metering to all homes by the end of 2020**. Smart meters let customers know exactly how much energy they are using and what they are spending on it, encouraging them to act on energy efficiency advice. They also mean no more bills based on estimates or staying at home for meter readings. Energy suppliers will be able to offer improved services, such as a wider range of tariffs and incentive packages, and the Government anticipates improved service levels and smoother switching between suppliers. Smart metering will support the use of microgeneration that in coming years, will help secure energy supply and provide a platform for future smart grids, which are described in chapter 3.

As an immediate step in this direction, *the Government has decided to include smart displays, which can be used with existing meters, amongst the range of measures that energy companies can offer their customers under the Carbon Emission Reduction Target.* As a result, the Government expects two to three million households to get a new tool to help them take control of their emissions, and see exactly how much energy they are using in real time. **To build on this and**

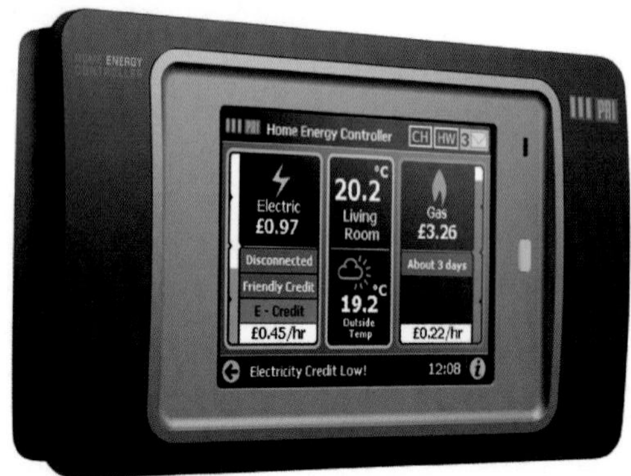

Smart meters and displays will help people to understand the energy they are using, and save money on bills

to reward people for doing the right thing, as well as helping people take action together, the Government will be launching a new national energy saving challenge this autumn, with prizes and incentives for saving carbon. The plans are being discussed with the energy companies, and further details will be available shortly.

The Government-funded Energy Saving Trust will develop more proactive services to make sure that the right advice reaches people at the right time and to help them get from the advice stage to implementation more easily. The Energy Saving Trust is already helping over three million customers a year through the Government's Act On CO_2 helpline, as well as a website and regional advice centres around the UK. This new service will involve:

- *Contacting all householders with properties that have received the lowest home energy ratings* (F and G ratings) in their Energy Performance Certificates, to offer tailored advice on energy savings steps that could be taken.
- *Providing a new Home Action Plan product,* where a household pays a subsidised cost for a thorough home energy audit, and a personal carbon reduction plan showing the energy saving steps to take, and their costs and savings potential.
- *Offering more support for people wanting to know how to generate their own low carbon energy.*
- Developing a Grant Information Database to *help people to find out about the funding offers available where they live.*
- Finding *ways to make it easier for people to find builders and tradespeople* with the right energy efficiency competencies

Helping people meet the costs of transformation

In the longer term, additional steps will be needed to overcome greater challenges such as bigger costs and disruption for householders, and there will be a need for greater levels of coordination across communities. New finance offers and different forms of financial support may be required to achieve this.

To test how to best help people to afford to make these changes, *the Government is now announcing that it will pilot a move from upfront payment to 'pay as you save' models of financing*, to help people to be able to meet the costs of transforming their homes to the best efficiency standards. This will spread the upfront costs into the future – the costs of improvements would be offset by energy bill savings. The Government will work with the Energy Saving Trust, energy companies, Local Authorities, the Distribution Network Operators (DNOs) and others to test uptake of this innovative financing, as well as householder interest in the 'whole house' approach. The Government will spend £4 million on these pilots and there is potential for match funding from a number of partners. Ofgem is considering scope for pilot funding for DNOs, under proposals for a new innovation incentive in the electricity distribution price control, to take effect from April 2010.

The Government wants to help people generate their own heat and electricity in low carbon ways, where appropriate. While these technologies typically save carbon and money, they can be more expensive to install, and so some financial support is necessary. To stimulate the UK's transition to local low carbon energy generation, **the Government will be providing more financial assistance to enable a much bigger roll-out, with 'clean energy cash back' schemes, including plans for the only stand-alone support scheme of its kind for renewable heat in the world**. The schemes include:

- **A new Renewable Heat Incentive (RHI) that will significantly ramp up the level of support available from April 2011**.[8] This will provide households, communities and businesses with payment for getting their heat from renewable sources. The scheme will cover industrial through to domestic scale heat production. *The Government will consult on the detailed design of the Renewable Heat Incentive later this year.*

7. The Government plans to consult on extending access to Energy Performance Certificate information so that Government Departments, local authorities, and relevant agencies are better able to target offers and support.
8. Renewable heat technologies include: air- and ground-source heat pumps, biomass fuelled stoves and boilers, solar-thermal water heaters and combined heat and power plants fuelled from renewable sources.

Figure 2:
Possible 'pay as you save' Model

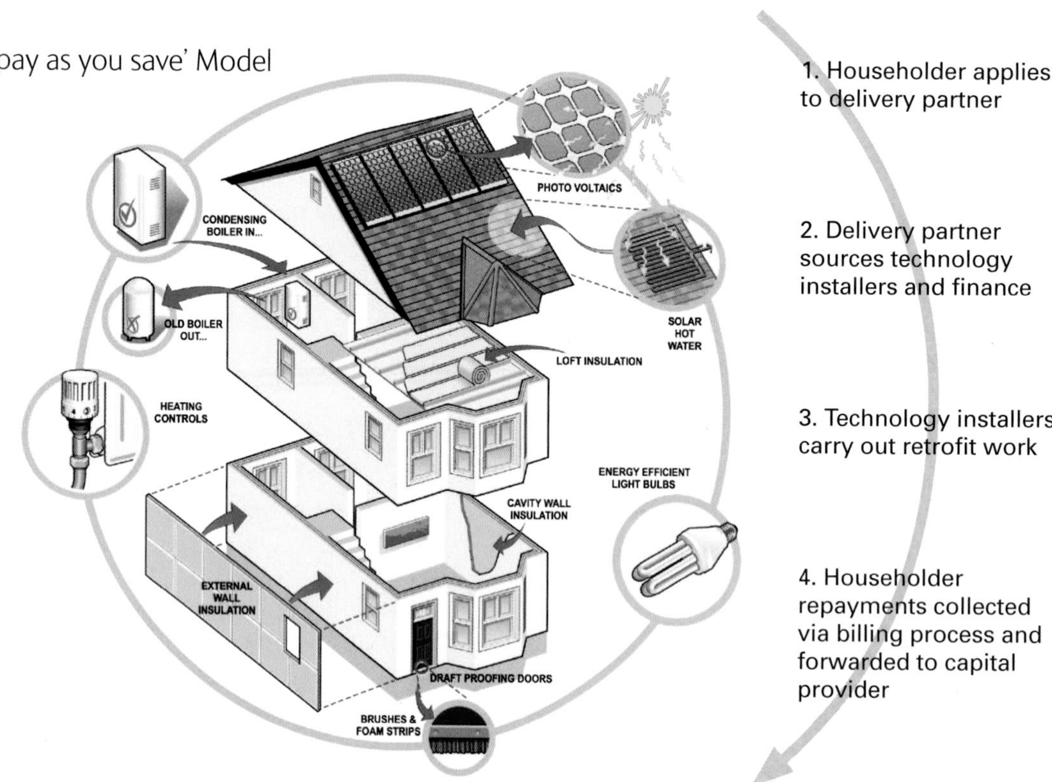

- **The Government is putting in place financial rewards for small-scale low carbon electricity generation, with Feed-in Tariffs from April 2010**. Payment for the electricity produced by small-scale generators, will be provided through the electricity supply companies and encourage the uptake of renewables by schools, homeowners, hospitals, businesses and communities.[9] *The Government is consulting on the detailed design and proposed tariff levels for Feed-in Tariffs alongside this Transition Plan.*[10] A household with a well-sited photovoltaic installation could receive over £800 plus bill savings of around £140 a year.

It is crucial that the RHI is available to everyone. The Government wants to make sure those on low incomes or who are fuel poor are able to benefit from this scheme and will be working closely with local authorities and community groups to help deliver renewable heat to fuel poor households. The *Renewable Energy Strategy*, published alongside this Transition Plan, gives more details of the two schemes, including the timeframes for operation.[11]

The UK's Renewable Heat Incentive will be the only scheme of its kind in the world, providing stand-alone support for renewable heat. Illustration of an air source heat pump.

9. See the Renewable Energy Strategy.
10. Department for Energy and Climate Change (2009) *Renewable Electricity Financial Incentives Consultation*
11. See Renewable Energy Strategy for details.

The increased burning of biomass in small scale boilers in areas with existing air quality problems may present an issue. Although the most modern domestic biomass boilers produce very low levels of air pollutants, their use in these areas needs to be carefully managed. The Government is considering whether, in the case of biomass boilers under 20MW, to limit the RHI to those boilers that meet certain emission standards, and will consult on this later this year. The Government has provided guidance to local authorities about what type of biomass boilers can be safely located where. The Government is also taking steps to ensure an increased supply of biomass can be met sustainably.

Co-ordinating the support available

Helping people to make the significant changes to their home required in the medium to long term will require strong co-ordination of the advice and funding support available – bringing together different aspects, such as energy saving measures and low-carbon energy generation. To enable installation of packages of measures across whole streets or communities rather than targeting individual houses, close links to local strategic decision making will be needed. As a result, a different approach to delivery may be required.

The Government is introducing a new community-based approach to delivering treatments to homes in low-income areas, the Community Energy Saving Programme, from Autumn 2009. This £350 million, three-year, programme will be funded by a new obligation on energy suppliers and electricity generators. This will see the energy companies working in partnership with local authorities and community organisations to deliver energy efficiency measures to around 90,000 homes across approximately 100 low-income areas across Great Britain.[12] This targeted programme of improvements will contribute to our carbon reduction objectives, support our fuel poverty strategy as discussed below, and inform future mechanisms for delivering energy saving measures.

The Government consulted in February 2009 on whether the current obligation on energy suppliers was fit for delivering energy savings in a more co-ordinated and 'whole house' way in the longer term.[13] A summary of responses has been published. It is generally felt that the supplier-led approach has been successful, but there was less consensus on the delivery approach best placed to deliver the scale of change now required. Some respondents were interested in the idea of greater central coordination, and some believed that Local Authorities were well placed to deliver. Other respondents were of the view that energy suppliers should continue to lead, and others favoured a combination of different players. **The Government is assessing the responses and intends to publish the strategy later in 2009**.

Raising standards in every home

As well as making it easier for people to transform their homes, the Government recognises that to enable everyone to make a contribution and to ensure the scale of change needed is achieved, we must raise standards in every type of housing.

Social sector housing is already more energy efficient than housing generally. *The Government has made major investment in improving social housing – more than £22 billion since 2001, including* through the Decent Homes programme. This has raised

12. The Government's response to its consultation on the Community Energy Saving Programme was published on 30 June 2009. http://www.decc.gov.uk/en/content/cms/consultations/open/cesp/cesp.aspx
13. Department for Energy and Climate Change *Heat and Energy Saving Strategy Consultation* (2009) http://hes.decc.gov.uk/

the condition of social housing, including warmth and comfort. By 2010, about 95% of social housing stock in England is expected to meet the Decent Homes standard, and following the last Budget, an additional £84 million will be invested in cavity wall insulation in English social housing. The Government will show leadership by ensuring that social housing meets, and where possible exceeds, the aims it is setting for all housing on energy efficiency and low carbon energy. *The Government is considering, with the Tenant Services Authority, the need to identify aspirational standards and benchmarks for energy savings and emissions reductions in refurbishment, for different property types.*

By 2010, around 95% of our social housing will meet the Decent Homes standard, meaning acceptable levels of comfort and warmth for tenants

Private rented accommodation is one of the most difficult sectors to improve because whilst it is the tenants that benefit from energy saving improvements as they pay the bills, it is landlords who own the property and decide what changes to make. To improve professionalism and quality in the sector, **the Government is consulting on whether to establish a register of private landlords, and whether to make the energy performance ratings of rented properties available to local authorities.**[14] These steps would build on the powers given to local authorities in 2004 to take enforcement action to improve the condition of rented housing – including energy efficiency standards – and would help to inform and target this work.[15]

In addition, the Government will:

- *Consult on requiring people to put Energy Performance Certificate (EPC) ratings for all rented properties alongside particulars of property advertisements*, so that potential tenants know more about energy performance from an early stage.
- *Consult on extending access to EPC information* for all homes so government departments, local authorities and relevant agencies are able to target offers and support.
- *Subject to final decisions to establish the landlords register proposed in the Rugg review, consult on giving the Energy Saving Trust access to that register* so that they are able to target offers and support to landlords more effectively, particularly identifying those with large numbers of properties.
- *Consider how to improve enforcement guidance to local authorities for the Housing Health and Safety Rating System*, in the light of experience of operating the regime.

14. Communities and Local Government (2009) *The private rented sector: professionalism and quality – The Government response to the Rugg Review*, http://www.communities.gov.uk/publications/housing/responseruggreview.
15. Under the Housing Act 2004, the Housing Health And Safety Rating System and associated enforcement powers for local authorities, ensure higher basic standards www.communities.gov.uk/documents/housing/pdf/1229922.pdf

The consultation on the long term Heat and Energy Saving Strategy set out a broad package of measures to help all homes make improvements. These will need time to bed down and become established before their impact can be properly assessed. **If these measures alone do not achieve the reductions in energy use required by 2012, the Government may need to consider stronger regulatory measures as part of its strategy. Ongoing evaluation will inform whether this is necessary**.

Looking to the future, we also need to make sure that our new homes and communities are built to high environmental standards. **New homes will be built to a zero carbon standard from 2016, meaning that their net carbon emissions over a year will be zero**. This will require high standards of energy efficiency, and any remaining carbon that is not abated onsite to be dealt with in ways to be specified. The Government has consulted on the detail of the definition of 'zero carbon', and will announce further thinking shortly. Progress towards the 'zero carbon' standard will be made through the progressive tightening of the Building Regulations and the Government recently set out proposals for the first step of 25% improvement in 2010.[16]

In 2007, the Code for Sustainable Homes became operational and homes are being completed at each of the levels.[17] The Code supports the zero carbon target and the changes in regulation along the way, allowing developers to be recognised for building to higher standards sooner.

Building zero carbon homes will require substantial change on the part of house builders and their suppliers. By setting the regulatory aims a long time in advance, and with defined steps, the Government aims to give certainty for investors and allow business to find the best way to meet the challenges. The Government is supporting the Zero Carbon Hub; to help home builders prepare through the Technology Strategy Board's Low Impact Buildings Innovation Platform it will assist business to harness the growing market for environmentally sustainable new buildings; and the Homes and Communities Agency is funding exemplar projects.[18] **Budget 2009 announced £100 million funding for local authorities to deliver new energy efficient homes, and *Building Britain's Future*[19] announced up to £250 million for direct development by local authorities of around 3,000 new energy efficient homes**.

New homes will be built to a zero carbon standard from 2016

Through the planning regime, and in particular the Planning Policy Statement: Planning and Climate Change,[20] the

16. Part L of the Building Regulations specify the minimum energy efficiency requirements for new buildings and certain categories of work to existing buildings in England and Wales. The current version came into effect in 2006, and improvements are proposed for 2010, 2013, and 2016.
17. The Code for Sustainable Homes is the Governments sustainable standard for house building, take into account energy, water, materials, ecology and flood prevention. www.communities.gov.uk/theCode
18. The Homes and Communities Agency is the national agency delivering homes and regeneration in England. Projects include the recently completed Code Level 6, Zero Carbon Homes in Northamptonshire.
19. HMG (2009) Building Britain's Future
20. http://www.communities.gov.uk/publications/planningandbuilding/ppsclimatechange

Box 2
The Olympic Park will be a blueprint for sustainable living

It is the Government's ambition that the London 2012 Olympic and Paralympic Games will be sustainable, with the Olympic Park a blueprint for sustainable living. It presents many opportunities to try new technologies, low carbon materials and innovative solutions.

Energy efficiency and water efficiency have been central to the design of the Olympic Park in East London. The aim is to reduce carbon emissions from the built environment in the Olympic Park by 50% by 2013. This will be achieved through:

- Making the Athletes' Village 44% more energy efficient than 2006 standards; and reducing water consumption by 20% compared to the London average.
- Using an innovative combined cooling, heat and power plant to supply the Park and Athletes' Village with energy, which will result in 20-25% reduction in carbon emissions in the longer term.
- Using on-site renewable energy sources during and after the games to reduce the call on conventional energy sources by 20%.
- Transporting 50% of construction materials by rail or river, and a target to reclaim 90% of demolition waste materials for reuse or recycling - this target continues to be exceeded.

In addition, the Park is being landscaped and planned to cope with climate change and flood risk.

Government requires new developments to be located and designed to reduce carbon emissions. The Government is also piloting new developments which meet the highest environmental standards on a large scale, for example with the eco-towns, the Thames Gateway eco-region and the London Olympic Park. These approaches will provide learning and help to point the way in decarbonising existing communities. *The Government will shortly publish its Planning Policy Statement: eco-towns, which sets out the high standards* that *eco-towns must achieve in tackling climate change and delivering affordable housing.*[21] Eco-towns offer opportunities to create thriving new communities, to test and bring to market new technologies, to support green skills and industries, and to help people to live more sustainably.

Helping communities to act together

Helping communities to take action is an integral part of the Government's strategy. We often achieve more acting together than as individuals. The role of the Government should be to create an environment where the innovation and ideas of communities can flourish, and people feel supported in making informed choices, so that living greener lives becomes easy and the norm.

In this vein, the Government has supported a number of initiatives. These include the Greener Living Fund, a £6.1 million fund for Third Sector delivery partners to help people live more sustainably; the Climate Challenge Fund, a £8.5 million fund to help communities

21. http://www.communities.gov.uk/housing/housingsupply/ecotowns/

Fifteen communities will be at the forefront of pioneering green initiatives with the new £10 million 'Green villages, towns and cities' challenge

raise awareness of climate change; and the Big Green Challenge from the National Endowment for Science, Technology and the Arts, a £1 million prize to galvanise community-led innovation in response to climate change. The Big Green Challenge is the first prize of its scale for the not-for-profit sector. Given the number and quality of the proposals, *the Government has now provided £600,000 to expand the Challenge and fund a new round of projects drawn from the top 100 unfunded proposals.*

The Government wants to take community transition to the next level, announcing £10 million for 'Green villages, towns and cities'– a challenge for communities to be at the forefront of pioneering green initiatives.[22] Around fifteen communities will be selected to participate as 'test hubs', with local residents, businesses, and the public sector playing a leading role, and eco-towns will be invited to participate. Participants will work together to develop community-wide plans

Box 3 Case study: Ashton Hayes used a grant from the Climate Challenge Fund to cut carbon emissions from the village

Ashton Hayes, in Cheshire, was the first village in the UK to announce its intention to become a 'zero carbon community'. Since January 2006, the community has cut its carbon dioxide emissions by 20%. The project has partnered with Chester University to establish the baseline and measure carbon savings. The village is currently working with partners to look into the feasibility of a 'microgrid' to generate and supply electricity to buildings in the village on an opt-in basis.

Ashton Hayes received support of £26,500 from the Government's Climate Challenge Fund, which was used to promote the initiative and share lessons with others. A film was produced and a conference was run for other community carbon projects. See www.goingcarbonneutral.co.uk

22. Funding is part of the low carbon investment funding announced in Budget 2009, see chapter 5 for more details.

for their neighbourhoods and learn how different initiatives – for example in energy and water conservation, or travel – work together in practice. They will be encouraged to explore new approaches to delivering to vulnerable groups by overcoming the particular barriers they face; and to share learning between the 'hubs' as part of a wider citizen-led pilot. If successful, the Government can use what we learn to help roll-out of a nationwide plan, potentially helping every city, town and village make the transition to a sustainable future.

Working with local and regional government

The local government performance framework gives an opportunity for local government and partners to reflect their priorities through the Local Area Agreements. Within this framework, around 97% of areas have already chosen to set targets against at least one of three climate change indicators. In addition, over 340 local authorities have signed the Nottingham Declaration on climate change.[23] Local government and its partners are clearly choosing to respond positively to the opportunity to agree targets on addressing climate change. **£3 million has been provided under the Best Practice Programme, which has helped local authorities take action by building capacity and spreading good practice.**[24]

The Government wants to encourage and empower local authorities to take additional action in tackling climate change, where they wish to do so. It believes that people should increasingly be able to look to their local authority not only to provide established services, but also to co-ordinate, tailor and drive the development of a low carbon economy in their area, in a way that suits their preferences. The Government will shortly consult stakeholders on the role that local authorities can play in meeting national carbon budgets, how this could work at a local level, and how this might combine an increased ambition on reducing carbon with possible new powers and flexibilities in this area.

The nine English regions are already taking action to help meet the UK's greenhouse gas targets and budgets. The Local Democracy Economic Development and Construction Bill will require each English region to develop a new single Regional Strategy, which must include plans to tackle climate change. It is essential that regional policy and action is consistent with the Government's emission reductions targets. Government Offices in the region will play an important role in supporting the delivery of this national strategy by working with partners at regional and local level.

Community-scale low carbon energy generation

As well as ensuring that governance frameworks are aligned, the Government wants to encourage local authorities and others in bringing forward more community-scale heat and electricity generation. For example, community heating provides 2% of heating needs in the UK, but it could play a bigger role of up to 14%.[25]

To help achieve an increase in community energy generation, **the Government has tasked local authorities to incorporate energy planning into their decision making processes, through the Climate**

23. The Nottingham Declaration commits the signatory authority to developing plans to address the causes and impacts of climate change according to local priorities.
24. Government funding through the Regional Improvement and Efficiency Partnerships.
25. Our analysis suggests that with appropriate changes in the market or policy arrangements, community heating could feasibly provide up to 14% of the UK's building heat demand. www.decc.gov.uk

Box 4
Birmingham saved over 100,000 tonnes of carbon dioxide last year.

Birmingham's Local Area Agreement set a target to save 100,000 tonnes of carbon dioxide between April 2008 to March 2009. Birmingham was successful in achieving this target - research showed a total saving of 103,039 tonnes. Initiatives that helped to reduce emissions included:

- Home insulation and low carbon programmes;
- Advice centres and community-based advisors offering information and practical help;
- Local activist groups raising awareness, such as the local Friends of the Earth group;
- The 'student switch off', a national energy-saving competition in halls of residence;
- Businesses cutting their emissions through changes in offices and staff travel;
- The City Council's work on highways and travel plans;
- Cleaner vehicles for West Midlands Police; and
- An increase in the amount of waste recycled, and a decrease in waste produced; and

More than two-thirds of the savings came from homes, 15.5% was from waste and water, and 12.5% from business and public sector.

Summerfield Eco Neighbourhood; and hydrogen fuel cell hybrid vehicles running at The University of Birmingham Campus

Change Planning Policy Statement. The Government also held a Local Authority Summit on Community Energy and Heating, in May 2009, bringing together leaders and key decision makers to share experiences from some of the more pioneering projects and authorities. **In Budget 2009, it announced £25 million to help to fund community heating infrastructure including at least 10 exemplar schemes across the UK**.[26]

To further encourage local authorities to strategically identify areas where community-scale infrastructure can be the most cost-effective solution, the Government, working closely with the Local Government Association, will:

- *Explore steps to encourage detailed heat mapping and planning by local authorities*, including through updating the Climate Change Planning Policy Statement.

26. Technologies being considered include biomass, energy from waste, deep geothermal and anaerobic digestion.

- *Examine the case for local authorities to have greater ability to require existing developments to connect to heating schemes*, to complement their ability to require new developments to connect.
- Investigate opportunities to strengthen the consideration given to CHP when companies are selecting the location and design of power *stations. The Government will review its guidance to developers on this with a view to ensuring both that developers give serious consideration to opportunities for Combined Heat and Power and identification of potential heat customers while at the same time making it easier for them to do so.*

To make it easier for local authorities, businesses or community groups to generate electricity at community scale the Government has worked with Ofgem to introduce new licensing arrangements that make it easier for community energy schemes to interact with the wider electricity system. Further work is planned to ensure arrangements work effectively in practice.

To offer further help, **the Government is now providing new funding to develop an online 'How to' guide for community energy, to be widely available from early 2010**. This will be an information hub for anyone looking to install renewable and low-carbon heat and electricity generating technologies at community scale. For example, it will highlight the benefits of cooperative ownership, and signpost community groups to existing information; and the Government will clarify the scope for local authorities to establish rolling investment funds, including through their own borrowing, designed to permit individuals and groups in their areas to fund schemes.

A new 'How to' guide will help anyone looking install renewable and low-carbon heat and electricity generating technologies at community scale

Protecting consumers

The alternative to meeting our carbon budgets is not a low cost, high carbon future, but a high cost high carbon future. We are becoming increasingly dependent on other countries for our oil and gas supplies. As set out in chapter 1, if the UK pursues a high carbon future, we will be exposed to global factors which could cause rises and spikes in energy prices. Last year, when oil prices reached nearly \$150/barrel, the cost of filling up a car tank rose £10 and electricity and gas bills rose £300 on average.

Although taking action on climate change is worth it, there will still be a cost. The Government has designed its policies to minimise these costs.

The measures set out in the Transition Plan will put upward pressure on energy prices and bills. The Government estimates that the additional impact in 2020 of policies in this Transition Plan, relative to today, is £76 per year, which is equivalent to approximately a 6% increase from current energy bills. When previously announced climate policies are included this figure is 8%.[1]

The increase in energy bills is caused by the use of renewable generating technologies

27. £76 as a proportion of average household energy bills today (approximately 1,184) is 6%. The 8% increase includes major recent climate policies, older/smaller policies are necessarily included in the baseline. Calculations assume \$80/barrel oil price in 2020. Bill impacts are dealt with in detail in the Analytical Annex to the Transition Plan.

that are more expensive than electricity from fossil fuels at today's prices. However, much of this increase is offset by the energy efficiency policies in the Transition Plan, including CERT, CESP, Warm Front, and smart meters policies.

Those who insulate their homes or take advantage of help to switch to renewable forms of heating will be partly shielded from these impacts. If oil prices reached their previous peak of around $150/barrel then the benefits of going green would be even greater. [28]

Keeping the price of energy competitive

We need our energy system to deliver results that are fair to consumers. For many years residential consumers have benefited from the UK's competitive energy market – in the period from 1997-2007, UK domestic gas prices have been among the lowest in the EU15.[29] Over the same period, UK domestic electricity prices have never been above the EU15 median. Consumers should also have the right information to enable them to make the right decisions about buying, using and saving energy.

Ofgem, the independent regulator, plays a key role in ensuring fair outcomes and treatment for consumers. In 2008, following concern from the Government and public about prices and other issues, Ofgem launched a detailed probe into retail energy markets. In its initial findings Ofgem reported that it had found no evidence of anti-competitive practices, but it did find evidence of unjustified pricing differences and other problems.

In response to Government and Ofgem concern at these findings, energy suppliers moved to address this unfair treatment and Ofgem is now concluding changes to outlaw such behaviour in future.

Ofgem's probe also found that energy bills and tariffs can be hard to understand. Following their recent consultation, Ofgem will very shortly be introducing licence modifications to enhance customer information via the bill and an annual statement. This will make it clearer to customers how much they are paying for their current tariff and make it easier for them to see if their supplier (or any other) offers anything cheaper for them. Ofgem has also decided to require suppliers to use a common price metric to further aid comparison of different suppliers' tariffs. This is a real priority which should greatly help consumers in making the right choices and controlling their energy bills. The Government has strongly supported Ofgem's work on the probe and where appropriate will act if the companies involved cannot agree. **The Government also proposes to strengthen consumer protection**, by:

- Enabling Ofgem to impose financial penalties for breaches of licence conditions which occurred earlier than in the previous twelve months (the current limit).
- Giving the regulator the ability to address cases of undue exploitation of market power in the generation market.
- Enabling the Secretary of State to outlaw certain types of cross-subsidy between gas and electricity supply businesses when these impact unfairly on groups of consumers, particularly electricity-only customers.

It is the Government's intention to legislate on these additional powers and enforcement matters as well as the clarification to Ofgem's principal objective at the earliest opportunity.

28. In addition, the Government's policies on minimum performance standards for appliances, will cut more than £30 per household or £900 million off householder bills per year.
29. EU15 are those countries that were part of the European Union before the 2004 enlargement. Price base includes tax.

Box 5
Clarifying Ofgem's remit

The Energy Act 2008 made it clearer that the principal objective of Ofgem in exercising its functions under the Gas Act 1986 and the Electricity Act 1989 is to protect the interests of future as well as existing consumers.

The Government now proposes to amend the legislation to clarify that this objective includes security of supply and reducing carbon emissions. These proposed changes will be complemented by new Social and Environmental Guidance to be published shortly.[30]

The Government continues to believe that effective competition remains the central way by which consumers' interests can be protected. However, there are contexts in which the promotion of competition may not be sufficient and direct action by the regulator is necessary to protect consumers effectively. The Government proposes to amend the legislation to make this clearer, building on the existing legislation. These changes will not undermine the central role that competition plays in the operation of the energy sector and the Government will engage with the sector before making them.

Protecting the most vulnerable

Tackling fuel poverty is a priority for the Government. The Government is under a statutory duty to end fuel poverty, as far as reasonably practicable, in vulnerable households by 2010 and in all other households by 2016 in England. Up until 2004, the number of households classed as "fuel poor" under the current definition of fuel poverty fell to a level of around 1.2 million households in England.[31] However, despite significant Government action and funding, the numbers of people living in fuel poverty have shifted adversely. The projected figures for 2008 suggest that the total number of households living in fuel poverty in England will have risen to just over 3.5 million households (see chart 6).

The number of people in fuel poverty depends on three factors: household energy

Chart 5

11% of UK households spent more than 10% of their income on energy in 2006

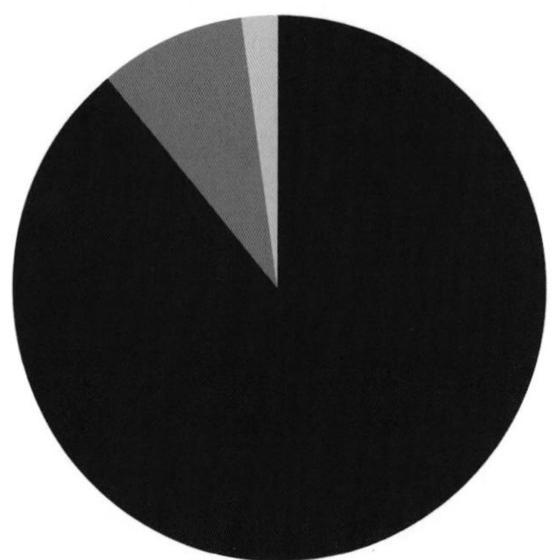

89% of households spend less than 10% of their income on energy bills

9% of households spend more than 10% of their income on energy bills and are categorised as vulnerable

2% of households spend more than 10% of income on energy bills and are categorised as non-vulnerable

Source: Department of Energy and Climate Change, *UK Fuel Poverty Strategy* (2008)

30. Under the Electricity Act 1989 and the Gas Act 1986, the Secretary of State must issue guidance from time to time about Ofgem's contribution towards the attainment of social or environmental policies to which Ofgem is required to have regard in the carrying out of its functions under the Act.
31. A household is said to be in fuel poverty if it needs to spend more than 10% of its income on fuel to maintain an adequate standard of warmth (usually defined as 21 °C for the main living area, and 18 °C for other occupied rooms). This broad definition of fuel costs also includes modeled spending on water heating, lights, appliances and cooking.

efficiency, household incomes and energy prices. One of the key drivers of the recent increase in fuel poverty has been rising energy bills due to higher fossil fuel prices: energy prices increased, on average, by 16% every year between 2004 and 2008. These significant price increases are making the fuel poverty targets ever more challenging.

Another key challenge in making further progress is being able to identify and target support effectively at fuel poor households, which requires information on the household's income, the price the household is paying for fuel, and the level of household energy efficiency. This picture is not static. For example: a household living in an energy inefficient home might not be in fuel poverty, but a lower income household moving into that same home might be; a household a little way out of fuel poverty might be brought into fuel poverty in the event of energy price increases; or a pensioner couple might not be in fuel poverty, but if one of them dies (and the household income consequently reduces) the other might be in fuel poverty.

For this reason, action to tackle fuel poverty is aimed not only at those people living in fuel poor households, but also to take preventative steps to stop households falling, or returning, into fuel poverty.

Since 2000, the Government has spent £20 billion on benefits and programmes to tackle fuel poverty. The Government's policies have been centred on:

- reducing the demand for energy through improving home energy efficiency;
- raising real incomes;
- ensuring competitive energy prices through regulating the market, and through voluntary social pricing support schemes.

Chart 6
The number of people living in fuel poverty fell until 2004

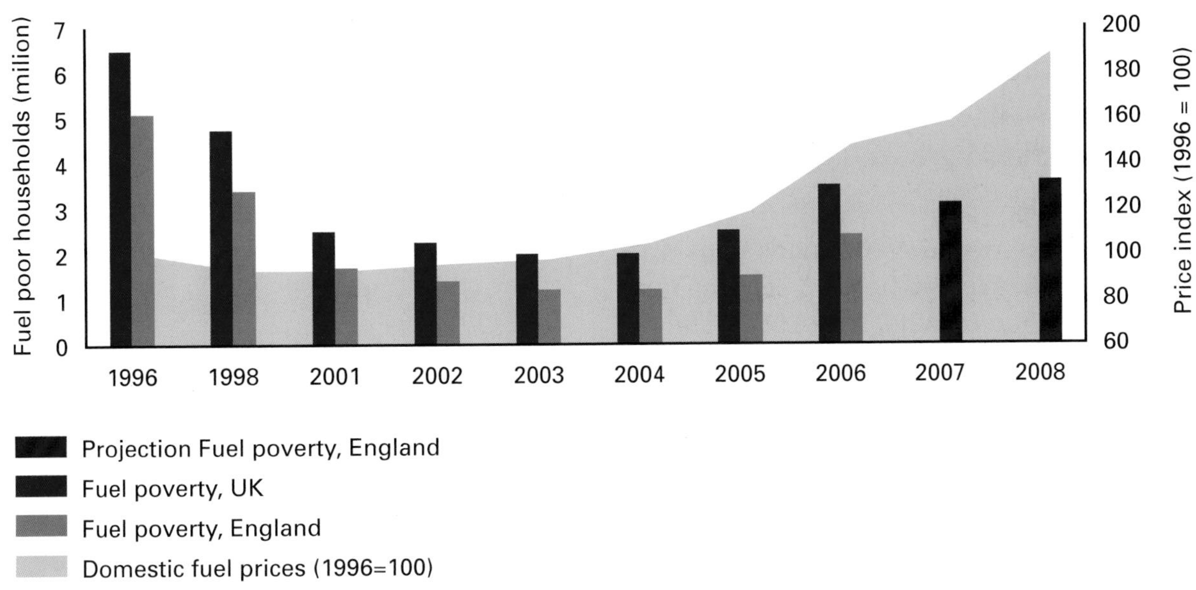

Source: Department of Energy and Climate Change (2008)

These measures have had a real impact. The Government estimates that without the package of measures aimed at tackling fuel poverty the number of fuel poor households would have been around 400,000-800,000 higher in England in 2008. The Government continues to look for ways in which it can make further progress. Earlier this year, the Government announced a review of its fuel poverty policies, examining whether existing measures to tackle fuel poverty could be made more effective, and considering whether new policies should be introduced.

This Transition Plan sets out existing policies, further progress made since the review was announced, and a new proposal on social pricing support. By 2020, the poorest households that are helped to install renewable heat technologies or insulation measures will have annual average energy bills £180 lower than if they had not received those measures.

Improve energy efficiency of vulnerable households

Some households need more energy to keep warm, as a consequence of the type of dwellings they live in, the heating systems used and of the characteristics of their lives (for example, how much time is spent at home). So improving household energy efficiency and the performance of heating systems are important ways of reducing energy bills for households in fuel poverty.

Warm Front is the Government's flagship energy efficiency and heating scheme for vulnerable households, with £950 million funding between 2008-2011. It has assisted nearly two million households since 2000– over half a million households in the last two years alone. On average, each recipient has the potential to save over £350 per year on energy bills.[32]

Box 6
Warm Front has helped almost two million households reduce their energy bills.

Mrs. Frost, 83, from Nottingham is in receipt of pension credit. She applied online to have her boiler replaced and two months later it was installed. Mrs Frost said: "I am delighted with the boiler replacement, everything was explained perfectly and everyone in the Warm Front Team were helpful".

In December 2008, Mrs. Whitbread, 77, from Preston had a new radiator and thermostat installed. Mrs Whitbread said, "I was very pleased with the service I received from the Warm Front Team; the installers were courteous, efficient and helpful."

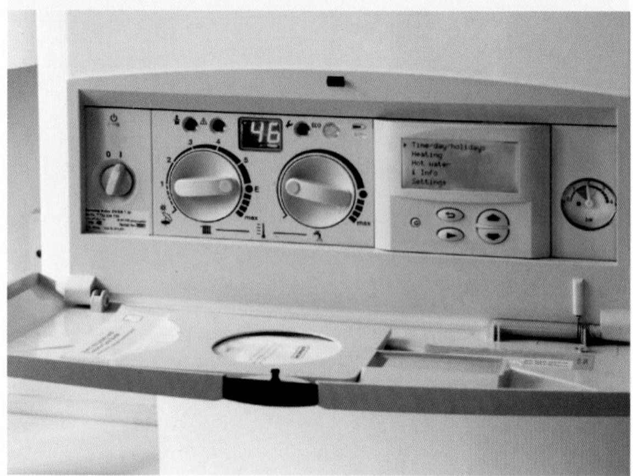

The Government is increasing the grant limits for eligible households to £3,500 (or £6,000 where oil or a new low carbon technology is recommended). This will mean that the vast majority of households will not have to contribute payment towards their measures.[33] The scheme is also being expanded to pilot the installation of low carbon technologies.

32. *Warm Front Annual Report* (2008-09)
33. In 2008 prices between 1996 and 2006

The new Community Energy Saving Programme (CESP), to be launched this autumn, is designed to be focused on areas of low income, where households are likely to have a greater propensity for entering fuel poverty than average. This new £350 million programme will improve energy efficiency and lower household fuel bills, and the partnership approach with local authorities and other community representative organisations, should help to reach some of the most vulnerable households.

Under the Carbon Emissions Reduction Target (CERT), energy suppliers have to achieve 40% of the emissions reduction in a priority group of low-income households and the elderly, who are more likely to be in danger of falling into fuel poverty. The obligation on suppliers is due to increase by 20% from August 2009, meaning an estimated £1.9 billion will be directed at energy savings amongst the priority group in the period to 2011. For the extension period of CERT post-March 2011, the Government is exploring how best to provide help to some of the most vulnerable households.

The Decent Homes programme in social housing aims to ensure homes are warm and weatherproof and have reasonably modern facilities. By 2010 work will have been completed to 3.6 million council homes, with improvements for eight million people in total, including 2.5 million children. The Decent Homes programme has on average reduced tenant's fuel bills by an estimated £152 a year.[34]

Raise incomes of the fuel poor

Raising incomes also helps to tackle fuel poverty. This can include employment measures, aimed at ensuring people are better equipped to work, such as Working Families Tax Credits; child benefit and child tax credit; increasing support for families on Income Support; the disability premium; and Pension Credit.

This winter, the Winter Fuel Payments and additional support will provide £250 for households with someone aged 60-79 and £400 for households with someone aged 80 and over (up from £20 when the Winter Fuel Payment was introduced in 1997). The Government also offers **Cold Weather Payments** for the most vulnerable people during very cold weather. Cold Weather Payments in Britain in 2008-09 amounted to around £210 million.

As mentioned above, some people entitled to benefits are not claiming them. Every applicant to the Warm Front scheme is offered a Benefit Entitlement Check, and since April 2008, 78,000 checks have been completed. Additional eligible benefit has been identified in 45% of cases, resulting in an average weekly increase in household income of £31.

There is also an important role for advice in protecting vulnerable households and the Government, in conjunction with other bodies, provides people over 60 with advice about how to stay healthy in the winter months.

Reducing energy prices for the most vulnerable

The measures taken by Ofgem to protect consumers also benefits people living in, or at risk of, fuel poverty, by keeping the price of energy competitive. To further help vulnerable customers unable to afford their bills, the Government has negotiated a voluntary agreement with energy suppliers to increase their expenditure on social programmes. Under this agreement, the suppliers delivered £100m of spend on social programmes to vulnerable households in 2008-9, rising to

34. The figures apply to applications made on or after 23 April 2009 and those applications made before that date but where the work has not commenced

£125 million in 2009-10 and £150 million in 2010-11. Examples of assistance from companies include rebates on annual bills and social tariffs. Ofgem estimated in December 2008 that 800,000 customer accounts were benefiting from some form of social or discounted tariff - almost double the number in March 2008.

The social price support available under the current voluntary agreement with energy suppliers has already made a real difference to the lives of a large number of vulnerable households. But the agreement comes to an end in March 2011. *The Government has decided to build on the success of the voluntary agreement and will therefore bring forward new legislation at the earliest opportunity with the aim of placing social price support on a statutory footing when the current voluntary agreement ends in March 2011*. As part of this new statutory framework, the Government will ensure there is an increase in the resources available and give the suppliers greater guidance and direction on the types of households eligible for future support.

The Government will continue to develop this policy over the coming months, engaging with interested parties and would expect to consult on the detailed arrangements of the scheme in 2010. Subject to further policy development, the Government is minded to focus a large part of the additional resources on those older pensioner households on the lowest incomes who are at greatest risk of excess winter death and who tend to have a high incidence of fuel poverty. The intention would be to offer support in the form of a fixed sum off the household electricity bill. This offers a clear benefit to recipients including those off the gas grid, minimises any distortion to competitive energy markets and maintains the incentive to make prudent use of energy. The Government will continue to work with the energy suppliers to explore more effective ways of targeting the price support at the most vulnerable households.

Ofgem has a role in administering and monitoring the current voluntary agreement. The Government is working with Ofgem to define Ofgem's continued role in this area.

As described earlier, the Government is also working to ensure that fuel poor households can benefit from new schemes such as the Renewable Heat Incentive, and the 'Green villages, towns and cities' communities scheme.

Formalising the current voluntary agreement will allow energy suppliers to extend price support to more of the most vulnerable households

Keeping gas supplies secure in the transition

The UK is heavily dependent on gas to heat our homes and provide domestic hot water. Around a third of UK primary gas consumption is used for this purpose.[35] Gas is also used by industry and for electricity generation (as set out in chapter 3).

In the long-term the UK must reduce its dependence on fossil fuels. As set out above, the Government plans to do this through reducing our need for gas through improving the energy efficiency of new and existing homes and by using alternative, low carbon sources of heat where available. In the medium-term the UK can begin to decarbonise its gas supplies by injecting renewable gas into the grid. The Renewable Energy Strategy sets out in more detail how renewable gas can be produced.

The Government expects that the measures outlined in this Transition Plan will reduce net UK gas demand by 29% in 2020, when it is expected the UK will be using around 66 billion cubic metres a year, compared to a demand without the Transition Plan of around 93 billion cubic metres a year. As a switch away from gas is made, there may be some scope for using parts of the gas transmission and distribution networks for other purposes, such as carbon capture storage.

But during the transition to a low carbon future, oil and gas will remain key sources of energy both in the UK and internationally. In the UK oil and gas currently supply 75% of our primary energy needs. The UK must be able to access reliable supplies to support the transition by avoiding interruptions and reducing the risk of avoidable price rises. The Government's approach includes:

- maximising the economic production of UK oil and gas – a large proportion of our gas is produced from oil fields;
- improving our capacity to import and to store gas;
- having in place the strategic partnerships to source gas imports.

Maximising the economic production of North Sea oil and gas

Realising the full potential of the UK's indigenous oil and gas resources, both offshore and onshore, is central to our security of supply, as well as having substantial benefits for the UK economy and employment – (taking account of supply-chain exports to the global market and other indirect employment, oil and gas activities support more than 400,000 jobs in the UK). But after four decades of oil and gas extraction, production is inevitably in decline.

Chart 7 shows central projections, but the actual level of UK production will be depend on a number of factors, and crucially on the level of investment and the success of further exploration. So far, around 40 billion barrels of oil equivalent have been produced in the UK, and as much as 20 billion, or more, could still be produced. To obtain full benefit for the UK from the remaining resources, it is essential that the UK continues to attract substantial further investment in a context of fierce international competition, and to maintain the presence of oil companies with the skills to identify and exploit these opportunities.

The Government has acted to support the necessary levels of investment and activity, including:

- **encouraging new exploration**, including in areas already explored, by quicker turnover of licences and by offering as much territory as possible for exploration;

35. 60% of final gas use is for residential sector.

Chart 7
UK production of oil and gas is declining and our dependence on imports is increasing

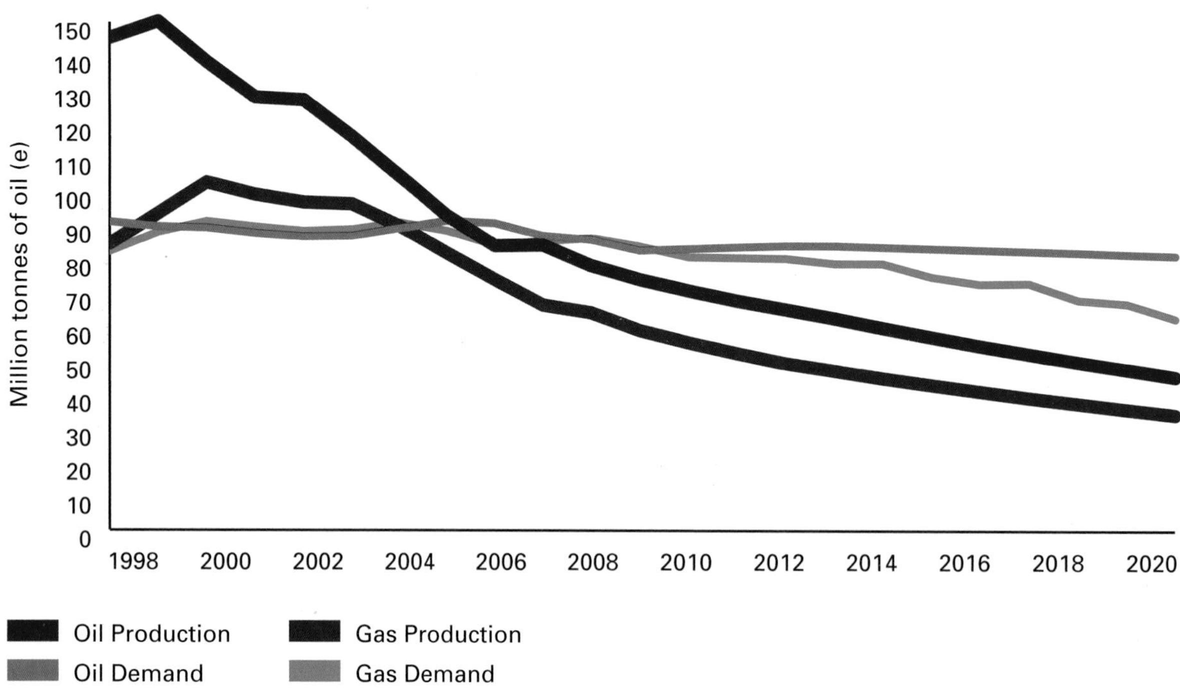

Source: Department of Energy and Climate Change (2009)

- **encouraging further development in existing fields**;
- **developing the tax regime** so that it continues to offer the right incentives for new investment.

The UK has been able to maintain continued strong interest from the industry in both offshore and onshore exploration and development. The last offshore round in 2008 resulted in the offer of 171 licences, the highest ever. The immediate prospects for investment are affected by higher levels of uncertainty stemming from the substantial fall in prices last year. But Budget 2009 introduced a new "Field Allowance" to give incentives for investment in small or technically challenging fields. The recently-published report of the House of Commons Energy and Climate Change Committee welcomed the introduction of the new allowance, and its analysis confirmed that it is likely to bring forward a significant tranche of projects. It nevertheless expressed concern that investment would not be sufficiently high without wider reforms of the fiscal regime. The Government will be considering that issue, alongside other actions which might help to sustain activity levels and skilled employment.

The area to the west of Shetland contains much of the resources not yet exploited. But new gas infrastructure is key to further development of the area, and until recently was lacking. However, the first major gas development in the area now seems likely, following the discovery of the Tormore field. On current planning, new infrastructure able to deliver gas to market could be available in 2014.

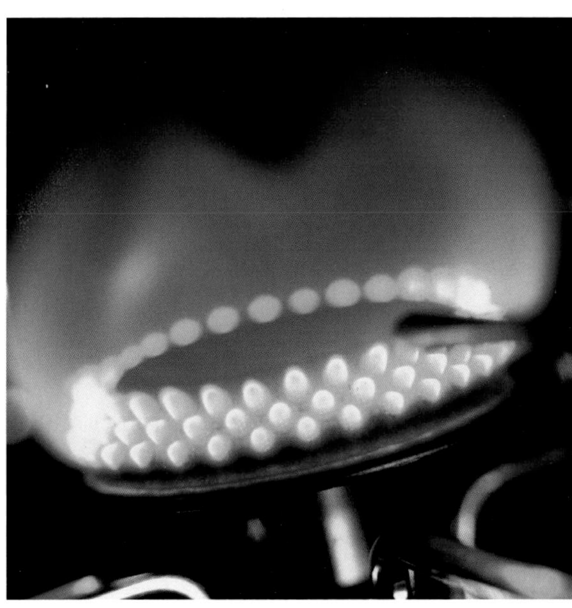

By 2020, indigenous gas supplies are likely to meet around half of our annual gas needs

International strategic partnerships

The UK has imported gas for many years and will continue to do so.[36] The diversity of our gas supplies helped the UK to remain largely unaffected by the Russia–Ukraine dispute in January 2009 (see box 7).

With domestic supplies declining and imports forming a greater proportion of overall gas supplies, the UK has a strong interest in improving the functioning of the global gas markets. Currently, most gas is traded bilaterally on long-term contracts and transported via pipelines. Gas can also be transported in ships as a liquid, allowing supplies to come from a greater variety of sources, and the liquefied natural gas (LNG) market is growing every year.[37]

The Government engages with a wide variety of countries, bilaterally and through the EU, to establish international strategic partnerships, and to provide the conditions for diverse and reliable gas supplies to the UK. For example, it works with Norway, Qatar, Turkmenistan, Nigeria and Algeria among others.

The Government also works multilaterally through the EU and other international bodies, including the International Energy Agency and the International Energy Forum to encourage the development of the global gas market. Issues that are particularly important include future investment in exploration and production of gas reserves, market transparency and functioning, and future technological developments, for instance those to exploit unconventional sources of gas.

As part of the European Union, the UK Government is taking steps together with its EU partners to improve how the EU's internal gas market functions, which will in turn strengthen security of supply in the UK and across the EU. The recently agreed "Third Package" of legislation on EU electricity and gas markets, is a big step forward. It contains a range of measures to encourage investment and improve the regulatory framework so that energy markets can work more efficiently.

The EU will soon revise the existing Security of Gas Supply Directive to help manage potential disruptions to Europe's energy supplies in a more co-ordinated and effective way. Similarly, the EU is working to ensure that it has diversity of energy sources, types, and routes by which they can be imported. This is important as the amount of gas that Europe imports as a proportion of total consumption will rise significantly in coming years and much of this will come from one source: Russia. The EU has therefore prioritised the development of a Southern Corridor route to increase the diversity of Europe's energy supply by bringing gas from the Caspian, via Turkey, to the heart of Europe.

36. The UK imports gas from a range of countries including: the Netherlands, Norway, Qatar and Algeria.
37. Liquefied Natural Gas (LNG) is a flexible and convenient way to transport gas to/from areas beyond the reach of gas pipelines. The gas is cooled to -161°C in the producer state, which turns the gas into a liquid and has the effect of shrinking its volume to 1/600th of its gaseous form. It is then loaded onto specially built LNG tankers and shipped around the world. The receiving terminals store the gas in large cryogenic chambers (on par with the size of the Albert Hall) until it is needed, when it is regassified by warming it up and then sent into the national grid

Box 7
Our experience during the Russia-Ukraine dispute last winter, and arrangements for this winter

The Russia-Ukraine gas dispute last winter saw a reduction in supplies of Russian gas to Ukraine from 1 January 2009, which escalated into a total cut off in Russian gas passing through Ukraine to Europe from 7 to 20 January. This had a significant impact on a number of countries in Europe, with some declaring states of emergencies.

The UK's diverse sources of gas imports meant that we experienced little direct effect, as we import very little of our gas from Russia. However, the dispute led to pressure on supplies in Europe to meet demand, and to the UK exporting gas to continental Europe in response to rising prices. Our experience last winter, compared to experiences in the winter of 2005-06, showed that the diversity of our supply sources and routes, and the recent expansion of import infrastructure, have strengthened our position against unexpected supply shocks.

A longer disruption, particularly if combined with a prolonged period of cold weather or additional disruptions to supply, could have resulted in greater pressure on prices. Although the immediate risks to UK security of supply were extremely low, the dispute highlighted some scope for improving our gas security of supply arrangements, to help manage supply-side shocks.

Against this background, National Grid is proposing some changes to the market framework for this coming winter. These include improved information and safety arrangements for gas storage and improved incentives for gas shippers to deliver gas into the system during an emergency.

Following the Russia–Ukraine dispute and UK lobbying, the Spring European Council agreed to bring forward the proposed revision of the 2004 Gas Security of Supply Directive. This is aimed at increasing Member State resilience to supply shocks, and improving emergency response mechanisms across the EU.

Improving our capacity to import

Indigenous North Sea resources supported about three-quarters of UK gas demand in 2008. But by 2020 this might have fallen to around 50% of our net annual demand. And, at times of peak demand in winter, UK production will contribute a much smaller proportion of demand and we will depend heavily on both gas imports and storage. To help manage the risks of this increasing dependence, the UK is developing a mix of gas import infrastructure and storage. The UK has a highly liberalised gas market which relies on companies to identify and address problems and these have brought forward gas import facilities, through both pipelines and as LNG. The UK now has three major pipes bringing gas direct from Norway and two from the Continent, and LNG can be imported at three sites.[38]

The UK's gas import capacity has increased more than five-fold over the last decade as a result of private sector investment, and

38. The Isle of Grain, Teesside and Milford Haven (where there are two separate terminals).

Chart 8

The UK is developing a mix of gas import infrastructure and storage, and the medium term outlook is encouraging

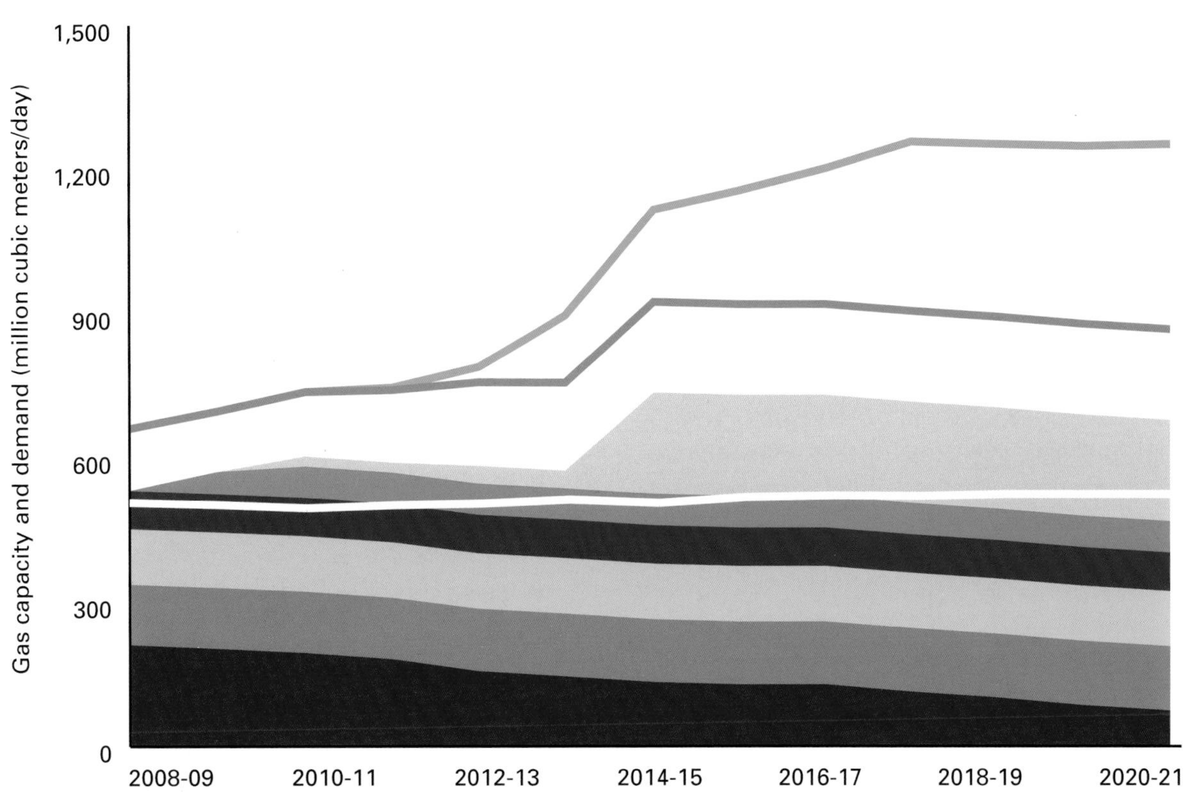

UK production (including biogas)
Import capacity of pipelines from Norway
Import capacity of pipelines from the Continent
Existing liquefied natural gas import capacity
Under construction liquefied natural gas import capacity
Possible new liquefied natural gas import capacity
Existing and under construction storage peak supply capacity
Possible new storage peak supply capacity
Demand (level of demand likely to be exceeded in only one out of every 20 years)

Source: National Grid (2009)

more capacity is under construction. The Government introduced the 2008 Planning Act to reform planning consents procedures, and this will enable the timely development of onshore gas storage projects, and the 2008 Energy Act paves the way for a consents regime for offshore proposals, such as storage and unloading facilities. The new tax relief for cushion gas (required to provide pressure within gas storage facilities[39]) will provide a further incentive to invest in storage capacity.[40]

39. Most gas storage facilities, whether in partially depleted petroleum fields or in salt caverns, require "cushion gas" to provide a "spring" to enable stored gas to be withdrawn at an acceptable rate and pressure when gas is being taken out of the facility.

40. The decision alongside Budget 2009, to make it eligible for tax relief through the capital allowances regime.

The outlook for our gas supplies in the short and medium-term is looking relatively positive. The recent expansion of import infrastructure and the diversity of gas supplies means that there is excess capacity to meet demand during severe weather conditions, in all but the most extreme instances of supply failure (see chart 8).

Further action

However, there are some risks that, because of their size, rarity and unpredictability, reduce the incentive for commercial operators to insure against them - although the public might expect otherwise. In the light of these, the Government will continue to take action to maintain the security of our gas supplies, by:

- **Ensuring the regulatory and fiscal regimes for the North Sea attract and support** the exploration and investment required to secure maximum benefit from the UK's hydrocarbon resources.
- **Facilitating the construction of new gas import and storage infrastructure**, in particular through reforming the consents regime.
- *The Government will shortly issue a commentary on the UK's ability to cope with severe demand and major gas supply shocks to 2025.*
- Malcolm Wicks MP has reviewed how the UK can maintain secure energy supplies during the transition to a low-carbon economy; his report will also be published soon.

The Government is consulting on legislation to substantially strengthen local flood risk management

Preparing our homes and communities for the changing climate

Using the planning system

The Government has built into the planning system clear expectations on adaptation to climate change: national policy statements on nationally significant infrastructure projects, regional strategies and local development documents must all take account of a changing climate. The Government has updated planning policies so that regional and local planners use information about our future climate to deliver planning strategies that secure new development in ways that minimise vulnerability and provide resilience to climate change.

The Government is consulting on legislation to substantially strengthen local flood risk management and make it a legal requirement to prepare flood risk assessments taking account of the latest climate projections. The Government has supported Sir Michael Pitt's recommendations following the 2007 floods. This includes additional funding to support the development of surface water management plans in high priority areas. The Government will consult on improving delivery of our planning policies on flooding.

Making buildings resilient

Our current building standards already help to ensure that our buildings are able to cope with future climate change, but the Government will also be considering the latest climate change projections as part of its on-going programme of review.

The Government has already made it a requirement on the builder to consider heat gains as well as heat losses in domestic buildings, and new minimum water efficiency standards for all new homes will come into force in October 2009.[41] Through the planning system, regulation, public funding and voluntary standards (such as the Code for Sustainable Homes), the Government is promoting strategies for adapted buildings–including green space and shading; passive ventilation; reflective glazing; water efficiency; sustainable drainage of rainwater; and water re-use.

41. Building Regulations, Part G and Part L

Chapter 5
Transforming our workplaces and jobs

Summary

The Transition Plan to 2020 involves two kinds of fundamental change in our workplaces:

- Reducing emissions: The energy we use in our workplaces accounted for 12% of UK greenhouse gas emissions in 2008.[1] To reduce emissions, we will need to change the way we do business to cut down the amount of energy and other resources we use, potentially saving businesses billions of pounds.
- New business opportunities: Many more of us will find ourselves working in a growing low carbon industry. Already 880,000 people in the UK work in the low carbon and environmental sector, a rapidly growing worldwide market worth £3 trillion per year and £106 billion per year in the UK. By 2020, this could rise to more than a million people if we seize the opportunity to establish the UK as a global centre of low carbon industries and green manufacturing. Around 200,000 of these new jobs by 2015 are expected to be in renewable energy, which could grow by a further 300,000 additional renewables jobs by 2020 as set out in the UK Renewable Energy Strategy, a total of half a million additional UK jobs in the renewable energy industry to 2020.[2] In doing this, the UK will need to focus on low carbon sectors

Chart 1

The workplaces sector will contribute about 9% of additional savings in 2018-22[3]

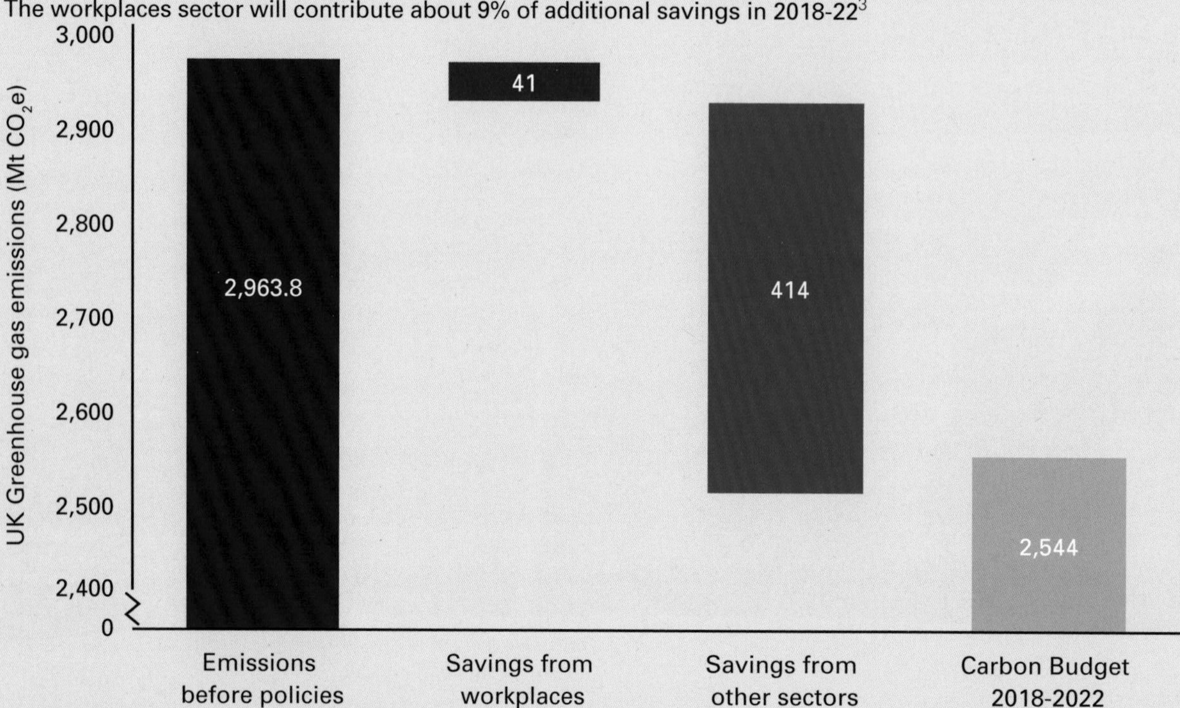

Note: Reductions due to policies introduced prior to the Energy White Paper 2007 are not shown. This chart also excludes the important role of the EU Emissions Trading System, which is set out in chapter 3.

Source: Department of Energy and Climate Change

1. Not including emissions covered by the EU ETS, as covered in chapter 3.
2. Department of Energy and Climate Change analysis
3. The emission and emission saving estimates in this chapter refer to greenhouse gas emissions from combustion of (gas, oil, and coal) in business and the public sector buildings where this is not covered by the EU Emissions Trading System (EU ETS). This will include the use of fuel in non-domestic buildings for heating and cooling as well fuel use for industrial processes where the industrial installations are below the EU ETS threshold.

where we are likely to have a competitive advantage such as offshore wind, marine energy, civil nuclear power, carbon capture and storage, renewable chemicals, low carbon construction and ultra-low carbon vehicles, and specialist financial and business services.

This Transition Plan, along with wider policies, will cut emissions from our workplaces by 13% on 2008 levels; help businesses grasp the opportunities arising from the transition to a low carbon economy; manage the costs associated with making this change; and secure the energy businesses need (see chapter 3).

The Government will help reduce emissions from workplaces by:

- Including high-carbon industries in the EU Emissions Trading System.
- Incentivising business and public sector workplaces to save energy through measures such as the Climate Change Levy, Climate Change Agreements and the Carbon Reduction Commitment.
- Providing advice for all workplaces on how to cut their carbon emissions and reduce their resource use.

To help make the UK a world centre of the green economy the Government is:

Workplaces contribute to climate change through the energy used to heat and light them and run equipment

- Investing in the low carbon industries of the future, including by using the £405 million of funding for low carbon investment announced in April 2009 to deliver a major boost to technologies where the UK has great potential, as described in more detail in the *UK Low Carbon Industrial Strategy*, published in parallel with this Transition Plan. This includes up to £120 million of investment in offshore wind, and investment of up to an additional £60 million to cement the UK's position as a global leader in marine energy, and help develop the South West of England as the UK's first Low Carbon Economic Area.
- Using other elements of the £405 million of funding for low carbon investment to deliver up to a further £10 million of support for ultra-low carbon vehicle infrastructure. The Government will also use up to £6 million to accelerate our progress towards a smart electrical grid and up to £6 million to support exploration of deep geothermal power – a new and innovative form of renewable energy.
- Supporting businesses through the global financial crisis and facilitating access to up to £4 billion of new capital for renewable and other energy projects from the European Investment Bank.
- Helping workers to develop the skills needed to thrive in a low carbon economy.

To make sure the transition happens in a fair way, the Government is:

- Providing loans and grants to help small- and medium-sized businesses and the public sector to afford upfront costs of those energy-efficient and low carbon technologies that will bring their energy bills down.
- Minimising impacts for all types of business by ensuring markets are competitive, providing a range of support and incentives for business to become more efficient and acting to keep EU frameworks fair.

The scale of the task

The transition will involve both reducing emissions from workplaces, and creating new jobs through maximising the business opportunities from low carbon goods and services. Around a sixth of the UK's greenhouse gas emissions are generated as a result of heating our workplaces, in addition to which, some businesses, such as steel makers, also burn fossil fuels as part of their industrial processes.[4]

The global industry in low carbon and environmental goods and services is already worth £3 trillion a year, and could grow to well over £4 trillion by the middle of the next decade. This offers an opportunity for the UK to capitalise on its advantages and establish itself as a world centre of green manufacturing. The UK low carbon sector is worth £106 billion a year – 3.5% of the global green sector – and employs 880,000 people.[5] A clear strategy for the development of low carbon industry across the UK's economy is vital if we are to realise the benefits this will bring, and preserve and enhance our competitive position internationally.

All workplaces will need to play their part in this transformation, but reducing emissions will be difficult because of a number of challenges:

- Reducing emissions is not seen as a strategic priority for many organisations, and many businesses and public sector organisations do not yet understand how they need to change.
- There is a big variety in the age and condition of the buildings we work in, and so a variety of solutions are needed.

Chart 2

Clean energy cash-back for heat will play a major role in reducing emissions from workplaces

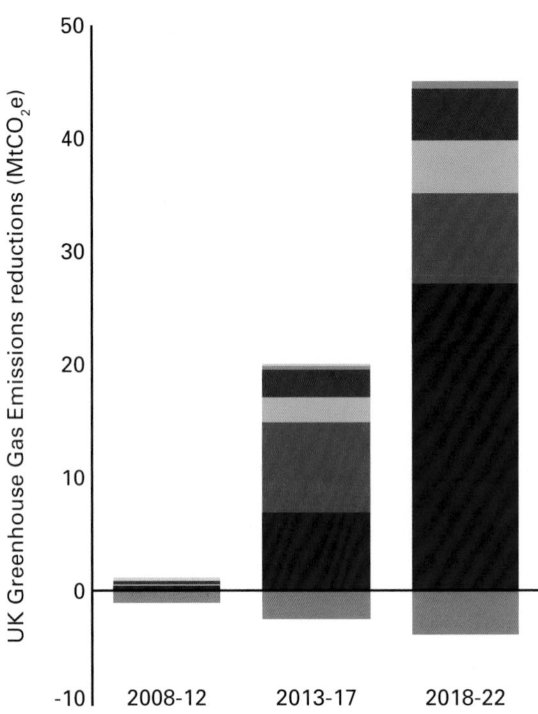

- Clean energy cash-back (renewable heat incentive) and supporting measures
- Energy intensive business package (Extension to Climate Change Agreements, Energy Performance of Buildings Directive, Carbon Trust advice and loans)
- Smart metering for small/medium sized enterprises
- Non-energy intensive business and public sector package (CRC, public sector targets, Carbon Trust loans)
- Energy Performance of Buildings Directive for small/medium sized enterprises
- Public sector loans including loans for small/medium sized enterprises
- Product policy for small/medium sized enterprises (see note to chart 4 in chapter four).

Note: Reductions due to policies introduced prior to the *Energy White Paper 2007* are not shown. This chart also excludes the important role of the EU Emissions Trading System, which is set out in chapter 3.

Source: Department of Energy and Climate Change

4. Businesses' use of electricity also contributes to emissions; the transformation of the power sector is covered in chapter 3.
5. Innovas (2009)

Chart 3

The global green economy is likely to expand significantly in the next few years

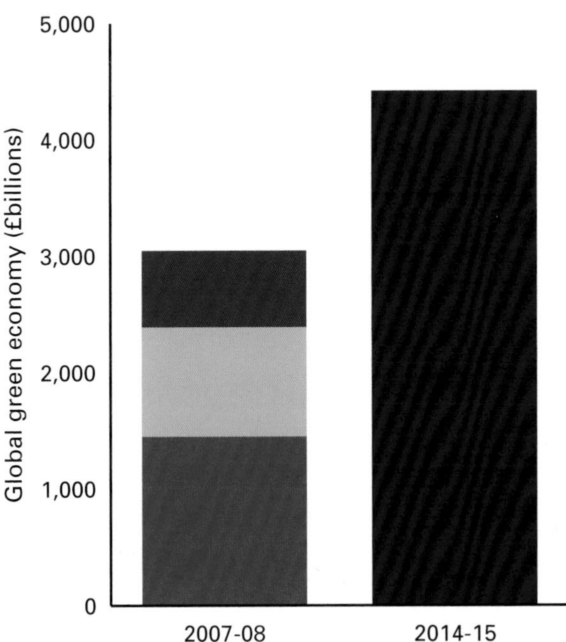

Potential size of global green economy by 2014-15

Emerging low carbon economy sector

Renewable energy sector

Environmental services sector

Source: Innovas (2009)

- Until technologies are familiar and proven, people and businesses are often understandably reluctant to use them, or are unaware how well-developed, robust and cost-effective 'new' technologies have become.
- Finding money to pay for new technologies is difficult, even for energy and resource efficient technologies where businesses get the money back quickly.
- Many businesses rent their premises and landlords can lack incentives to make changes.

880,000 people work in the UK green sector

Businesses wishing to develop and sell new low carbon technologies also face challenges:

- Getting access to long term capital can be difficult because the longer payback periods and risk of technology failure can put businesses and investors off, especially with the current constraints in the credit market. This challenge is particularly difficult for small- and medium-sized businesses with limited or no track record.
- Other countries are also moving to take advantage of new low carbon opportunities and there will be global competition to lead the future low carbon industries.
- Making the most of new opportunities will require employees to develop new skills.

Low carbon technologies like this ground source heat system can be a good way of heating our workplaces

The plan to 2020

The policies set out in this plan will ensure **emissions from our workplaces are cut by 13% on 2008 levels**, and help businesses capture the opportunities from the move to a low carbon economy, whilst keeping the financial impact on businesses down.

This Transition Plan sets out how the Government will act to help every business manage the costs of making the change to low carbon while taking up its opportunities. There is a real chance for businesses to gain in two ways.

First, businesses can realise significant and long-term savings by reducing emissions from their workplaces. If all UK business undertook cost effective measures then collectively £6.4 billion could be saved – 2% of UK profits. The Government will support business by:

- **helping businesses to reduce emissions, and**
- **taking action to lead the way in the public sector on emissions savings.**

Second, the growth of the £3 trillion global low carbon industry offers the chance for our businesses to invest in new opportunities and create new jobs. The *UK Low Carbon Industrial Strategy*, published in parallel with this document, sets out in more detail how the Government will work alongside business to make sure the UK is well placed to succeed as a centre of green manufacturing. The Government will help business to benefit from this growing international marketplace by:

- **Grasping the benefits of new economic opportunities, and**
- **Helping to manage the costs of making the transition to a low carbon world**

Alongside these opportunities, it will be essential that UK businesses are able to rely on secure gas and other energy supplies as set out in Chapters numbers.

Reducing emissions from our workplaces

Helping businesses to reduce emissions

Introducing caps and incentives for businesses to reduce emissions

The Government is committed to working with businesses of all sizes to make the emissions reductions that all of our workplaces need to achieve. Underpinning the Government's strategy for larger, more energy intensive businesses is the **EU Emissions Trading System (EU ETS)**, which since 2005 has put a cap on about half of the UK's CO_2 emissions including those from more carbon intensive businesses such as those producing steel and cement. The system has already played a role in guiding business investment decisions. By 2020, it will guarantee annual emissions reductions of around 500 million tonnes of CO_2 across the EU when compared to 2005, and there will be further reductions once there is a comprehensive international climate agreement. Further details about the EU ETS are set out in chapter 3.

Until there is an international agreement, the small number of sectors that are genuinely at significant risk of "carbon leakage"

Box 1
Shotton Paper Mill – a case study of investment in greenhouse gas emissions reduction under EU ETS

UPM is one of the world's leading paper producers with plants in 15 countries. It has two mills in the UK which are covered by the EU ETS. The Shotton paper mill in North Wales, which produces 500,000 tonnes per annum of 100% recycled content newsprint, has invested £52 million in a biomass Combined Heat & Power (CHP) plant. This plant has reduced electricity imports from the grid by almost a third and virtually eliminated the use of gas and gas oil, leading to a saving of 100,000 tonnes of CO_2 per year.

Shotton paper mill is saving CO_2 with a biomass Combined Heat and Power plant

(business going overseas to avoid restrictions on greenhouse gas emissions) will receive free allowances under the EU ETS. These free allowances will be given on the basis of stretching performance benchmarks set at the level of the most carbon-efficient businesses in each sector.

The Government is also helping many other businesses which, while large, are less energy intensive than those covered by the EU ETS, to reduce their emissions by introducing the **Carbon Reduction Commitment** to drive a reduction in their energy consumption. This new scheme, which will begin in April 2010, will require large non-energy intensive organisations to purchase carbon allowances, the total amount of which will be limited from 2013 onwards, to cover their energy use emissions. Revenue raised from the sale of allowances will then be recycled to participants according to how they perform in reducing emissions. An annual league table ranking the performance of all participants will also be published.

The scheme is expected to deliver substantial cost savings of around £1 billion by 2020 due to savings in energy bills. And we expect carbon savings to be at least 4.4$MtCO_2$. The level of the cap on emissions under the Carbon Reduction Commitment is expected to be set in 2012 following advice received from the Committee on Climate Change in 2010, and some experience of the scheme's operation.

The Government has also introduced the **Climate Change Levy (CCL)** to encourage businesses across the economy to use energy more efficiently and **Climate Change Agreements (CCAs)** to incentivise energy intensive businesses to take action while remaining competitive.

The Climate Change Levy is an energy tax, the aim of which is to encourage businesses to use energy more efficiently. It is charged on energy supplied to business and the public sector, but not, for example, on renewables or on good quality combined heat and power plants. Fuel supplied for electricity generation and most fuels supplied for transport are also excluded.

Climate Change Agreements were established to mitigate the impact of the Levy on the competitiveness of energy intensive industry, whilst also securing uptake of energy efficiency opportunities. CCAs are voluntary agreements between government and industry that enable eligible energy intensive businesses to obtain an 80% discount from the CCL in return for meeting challenging, but cost effective, energy efficiency or carbon saving targets.

The Climate Change Agreements and Levy are estimated to save around 23 million tonnes of carbon dioxide a year by 2010, despite strong output growth in many of the sectors involved, and businesses with Climate Change Agreements are currently estimated to save around £1.7 billion annually in energy costs when compared to baselines.

The current Climate Change Agreements end in 2013. Subject to State aid approval they will be extended to 2017. The Government is consulting on the form and content of the new agreements. Negotiation of targets with industry for the new CCAs has yet to start. However, the Government believes there is significant scope for additional cost effective emissions reduction by sectors currently covered by CCAs. We therefore intend to review the potential, to deliver at least 8 million tonnes of cost effective greenhouse gas abatement in the non-traded sector in the second and third carbon budget periods.[6]

Combined heat and power (CHP) plants, which provide more efficient, cost effective energy by harnessing the heat created as a by-product of power generation, are supported by exemptions from the Climate Change Levy. **The Government announced in Budget 2009 that the current exemption from the Climate Change Levy for good quality CHP electricity sold to the grid will be extended by a further 10 years to 2023**, subject to further state aid approval. It is estimated that this long-term market signal will help to unlock an additional £2.5 billion of investment in large-scale CHP projects.

Cement works near Stirling, Scotland: Climate Change Agreements are helping industry stay competitive

6. No decision has been made on the future of CCAs beyond 2017

CHP facility in Lewisham, South East London
Businesses are exempted from climate change Levy for good-quality combined heat and power plant

Currently, most refrigerators and air conditioners in commercial buildings and in all vehicles use fluorinated gases as the refrigerant. These make up almost 2% of UK emissions. To set a regulatory framework to enable businesses to reduce emissions from the products they sell, the Government has worked at European level to develop legislation to reduce EU/UK fluorinated gas emissions which has succeeded in reducing emissions by 39% since 1995.

Other action at EU level is also helping to reduce emissions, including the Energy Performance of Buildings Directive, which includes measures relating to Energy Performance Certificates, Display Energy Certificates for public buildings, inspections for air conditioning systems, and advice and guidance for boiler users. The Government is also helping businesses to save energy and resources in other ways such as proposals being explored through the Government's consultation on the roll-out of smart gas and electrical meters to households and certain smaller non-domestic customers. Budget 2008 announced an ambition for all new non-domestic buildings to be zero carbon from 2019, and for new public sector buildings to be zero carbon from 2018.[7]

Providing financial support

The Government helps businesses that would otherwise find it hard to pay for new technology, to get the finance upfront.

Box 2
The Carbon Trust is working to accelerate the move to a low carbon economy

The Government-funded Carbon Trust helps to cut carbon emissions by providing business and the public sector with expert advice, finance (including interest free loans) and accreditation, and by stimulating demand for low carbon products and services. Through this work, Carbon Trust helped save over 17 million tonnes of carbon, delivering costs savings of over £1 billion, between 2001 and 2007/08. Carbon Trust also helps cut future carbon emissions by developing new low carbon technologies through project funding and management, investment and collaboration and by identifying market failures and practical ways to overcome them.

7. The Definition of Zero Carbon Homes and Non-Domestic Buildings consultation, published on 17 December 2008, set out the Government's current thinking in this area. The Government is currently analysing the responses and will consult later in 2009 on more detailed proposals.

Box 3
Precious Little One used a Carbon Trust Energy Efficiency Loan to cut bills and emissions

Nursery retailer Precious Little One needed an efficient heating system that would pay its way in the long term, without adding to its immediate financial burden. An interest-free loan of £10,000 from the Carbon Trust allowed the company to install an energy-efficient replacement for its old heating system. Annual savings are estimated at over £5,000 making the payback period just two years, and saving an estimated 27 tonnes of greenhouse gas emissions every year.

Small and medium-sized businesses find securing financing particularly difficult, especially in the current economic context. In 2007/08 over 700 Government-backed interest free loans through the Carbon Trust helped businesses save around £9 million a year. **In April 2009 the Government announced a further £100 million of available funding to enable up to 2,600 more businesses in England to save money over the next two years.** Extra financial help is also available to a wider range of businesses investing in some of the 14,000-plus products which are eligible for the Enhanced Capital Allowances Scheme, helping them to realise further savings.

Government is helping business secure finance to make the transition to a low carbon world

But the Government realises that this will not roll out low carbon energy fast enough. So it is introducing a **"clean energy cash back"** scheme for workplaces that produce their own energy renewably (the feed-in tariff from 2010), and which install renewable heat technologies (the renewable heat incentive from 2011). Further details are set out in chapter 4 (Transforming our homes and communities).

Leading the way in the public sector

Central government and the wider public sector must lead the way in reducing their own carbon emissions and driving the move to a low carbon economy. The public sector is directly responsible for around 1% of UK's emissions. Public sector emissions have already reduced by a third between 1990 and

2007, compared to an 18% reduction by the UK economy as a whole. **The Government has plans in place that will deliver a total reduction of 16.9% across central Government offices by 2010/11,[8] against a target of 12.5%.** The Government's ambitions for the public sector will be reflected in the departmental carbon budgets described in chapter 2, and will incorporate emissions savings from tens of thousands of public institutions.

Government departments have been set a target of a 30% reduction in their own estate and operations emissions by 2020 from 1999 levels. *This target will be incorporated into departmental carbon budgets and exceeds that set for the economy as a whole (34%, but against 1990 levels).*

Action is being taken to save emissions across the public sector

The Government will be using up to £1.75 million of low carbon investment funding to support the creation of a flagship district heating scheme for London in partnership with the London Development Agency. The Government will fund the installation of a new heat main and combined heat and power plant, to create an integrated scheme with the potential to deliver annual savings of 10,000 tonnes of CO_2, and savings to heating bills for those connected. Adjacent Government departments and other public sector buildings will be encouraged to connect to the scheme.

Budget 2008 announced the Government's ambition for new public sector buildings to be zero carbon from 2018. In addition, action is taking place across the public sector:

- All new schools will be zero carbon by 2016 and the higher education sector is developing a carbon reduction strategy. Over the next fifteen years, all secondary schools and up to 50% of primary schools will be refurbished to be better adapted to climate change and have lower carbon footprints.
- The NHS plans to reduce the level of its 2007 emissions by 10% by 2015, and has developed a strategy to achieve this.
- 35 Local Authorities have committed to set targets in their Local Area Agreements to reduce greenhouse gas emissions from their operations, and all authorities will be required to report progress against these national indicators, with outcomes publicly reported from November 2009.

8. On 1999 baseline levels.

Box 4
Cutting the Department of Energy and Climate Change's emissions

After its creation in October 2008, the Department of Energy and Climate Change took over a G-rated Government building – the lowest rating available. The Department has enlisted the Carbon Trust's help to conduct a carbon survey of its buildings and develop a carbon management plan to progressively reduce annual emissions, making changes to optimise its heating, ventilation and cooling systems as well as its head office lighting using motion and daylight sensing. These are designed to achieve an ambitious goal of a reduction of 10% of the building's greenhouse gas emissions in the 2009/10 financial year, and further reductions beyond this.

The Department of Energy and Climate Change expects to cut energy bills

- Over 160 of the Government's 261 overseas diplomatic posts have taken action to reduce their environmental impact. Projects undertaken have been as diverse as connecting the Ambassador's Residence in Stockholm to district heating and piloting solar panels for water heating in Nairobi.

In addition to targets, emissions from all central Government departments, larger public sector organisations and state schools will be limited by inclusion in the forthcoming Carbon Reduction Commitment. The public sector is also incentivised by the climate change levy and CHP exemption set out above.

To ensure that the public sector leads the way and that it implements measures to reduce emissions rapidly and effectively, *the Department of Energy and Climate Change is working with HM Treasury to establish a cross-cutting review of energy efficiency in the public sector.*

Information and advice is just as important as setting targets and helps workers to think about reducing emissions. The Carbon Trust Public Sector Carbon Management service has helped 440 public sector bodies to identify savings of around £200 million. This means they can aim for ambitious carbon reduction targets of up to 28% over the next five years.

The Government will also provide targeted financial support where appropriate, and is considering plans to support a series of demonstration sites across the public sector estate and in major public buildings. These demonstration schemes will accelerate the deployment of ultra efficient lighting where the technology is ready – for example replacing halogen spotlights in cafes, atria, corridors and floodlighting with light emitting diodes.

Box 5
Salix saves Nottingham schools thousands of pounds in energy bills

Fountaindale School is saving around £11,000 a year on its heating bill from investing £7,800 to insulate the pipes in the school's heating system. Tollerton Primary School is saving £1,200 a year on its electricity bill from investing £4,210 to upgrade its lighting.

Making finance easy for public sector organisations

Just like private businesses, public sector bodies can have difficulty paying for energy efficient improvements up front. To address this, loans for the public sector are available through a scheme operated by Salix Finance. In April 2009 an additional £54.5 million was made available to, among other organisations, schools, leisure centres, Local Authorities and central Government departments in England. This will reduce public sector energy bills by around £14 million a year, showing energy savings can be both good for climate change and for public services.

In April 2009, the Government announced an additional £45 million to provide grants for small-scale renewable technologies through the Low Carbon Buildings Programme, helping homes, charities, businesses and the public sector to invest in renewable energy technologies.

Benefitting from the low carbon economy

Grasping the benefits of new economic opportunities

The global move to a low carbon economy is creating ever growing demand for low carbon technologies, goods and services. The Government wants the UK to be a global centre for the green manufacturing that will deliver this technology.The UK must be not only at the forefront of using renewable energy and other low carbon technologies, but also of building and exporting them, with the jobs and economic opportunities that go along with that. **For example, a recent study estimated that carbon capture and storage technology could bring between £2 and 4 billion a year into the UK economy by 2030, and support between 30,000 and 60,000 jobs.**[9]

Schools: like this one in Middlesex can benefit from green technologies to save thousands of pounds on energy bills.

9. AEA Group (2009) Future of Coal Carbon Abatement Technologies to UK Industry

The *UK Low Carbon Industrial Strategy*, published in parallel with this document, sets out the scale of potential business opportunities and a programme of Government action. The Government is acting to support business by putting in place a coherent framework of support and opportunity:

- **Providing targeted support for innovation and investment in the UK.**
- **Supporting businesses through the global financial crisis.**
- **Developing the skills needed for a low carbon economy.**

Targeted support for innovation and investment in the UK

UK businesses are in a strong position to lead the way in key sectors including offshore wind and marine energy, civil nuclear power, carbon capture and storage, low carbon construction, low carbon vehicles and specialist financial and business services such as green venture capital or environmental consultancy that support the low carbon sector. Information and communication technology will also be a critical enabler of low carbon activity across all these sectors.

Bringing new and innovative technology to market is essential not only for UK businesses to compete in the low carbon market, but also as part of the global effort to combat climate change effectively. Both domestic efforts and international collaboration are needed to develop the technology that will be fundamental to making a cost-effective change to a global low carbon economy.

The Government already works through international fora, including the IEA, G8, G20 and the Major Economies Forum to enhance the development and deployment of low carbon technologies, focussing on technologies that will produce the largest greenhouse gas reductions. This has involved working closely with the IEA on energy technology roadmaps, collaborative R&D, capacity building and funding of technology deployment.

Barriers exist, however, to UK businesses exploiting the country's potential advantages which the Government must help to overcome. Budget 2009 announced £405 million of new funding to help secure the UK's status as a global centre of low carbon industries. In *Investing in a Low Carbon Britain*, published on 23 April, the Government announced that this funding would be used to encourage businesses to invest in the UK through support for innovative low carbon technologies, and the development of our low carbon industry and its supply chain.

The funding supports a range of initiatives including support to communities and householders detailed in chapter 4 and to help the public sector to take a lead as set out earlier in this chapter. The *UK Low Carbon Industrial Strategy* sets out a range of activity that will be brought on by this funding with further announcements to follow. Current plans include:

Offshore wind:

- The Government has earmarked *up to £120 million to support a step-change in investment in the development of the offshore wind industry in the UK.* This includes funding for new offshore wind energy manufacturing facilities; investment in the development of next-generation and near-market offshore wind technologies through large scale demonstration; and improvements to the UK's capability in integrated offshore wind testing including through dedicated testing facilities.

Marine:

- *The Government is allocating up to an additional £60 million for a suite of measures which will help accelerate the development and deployment of wave and tidal energy in the UK* and will cement our current position as a global leader in the sector.
- *The Government will double its financial support to Wave Hub* – the development of a significant demonstration and testing facility off the Cornish coast – with up to £9.5 million of investment. *The Government is also proposing to invest up to £10 million at NaREC*, the New and Renewable Energy Centre, in the North East to build on and utilise existing infrastructure to provide an open access facility for marine developers to test and prove designs/components onshore. The Government will also provide *up to £10 million to support the South West's significant potential for wave and tidal energy deployment, research, demonstration and engineering, and up to an additional £8 million from the UK Environmental Transformation Fund to expand the in-sea stage testing facilities at EMEC*, the European Marine Energy Centre, in the Orkneys.
- In addition the Government will launch a *Marine Renewables Proving Fund which will provide up to £22 million of grant funding* for the testing and demonstration of pre-commercial wave and tidal stream devices. This will accelerate wave and tidal technologies' move towards commercial demonstration and assist the development of successful projects under the Marine Renewable Deployment Fund. Taken together, these investments will provide the UK with unparalleled testing and demonstration facilities.

Ultra-low carbon vehicles:

- The Government is providing up to *a further £10 million for the accelerated deployment of electric vehicles charging infrastructure in the UK. And will establish a new cross-Whitehall Office for Low Emission Vehicles (OLEV)* that will drive policy delivery. This will complement the £20 million for infrastructure, £140 million for research, development and demonstration under the Technology Strategy Board's Low Carbon Vehicle Innovation Platform, and £230 million for consumer incentives announced earlier this year.

A 'Smart' electrical grid

- *To accelerate the UK's progress towards a smart electrical grid capable of integrating generation and consumption of energy more effectively than ever before, the Government will provide up to £6 million* to complement other funding for network innovation such as Ofgem's Innovation Funding Initiative amongst other sources. Government funding for smart grids will be used to support early stage development of trials of key technologies consistent with a vision for smart grid in the UK to be published later in 2009 (See chapter 3 for further details on the development of a smart grid).

Low carbon construction

- *Renewable resources can substitute for oil based materials in a wide range of other non energy applications, for instance construction products, industrial chemicals, and plastics. The Government is investing up to £6 million to construct 60 more low carbon affordable homes* built with innovative, highly insulating, renewable materials. The new scheme will demonstrate the viability of these materials, and act as a spur for the renewable construction materials industry. It will also and to engage the affordable housing sector in the low carbon agenda.

Deep geothermal power

- Deep geothermal power is an innovative energy technology that is seeing a surge in interest worldwide. It uses the natural heat from deep underground to drive turbines at the surface, providing a renewable and non-intermittent source of electricity and heat. *The Government will commit up to £6 million to explore the potential for deep geothermal power in the UK*, helping companies carry out exploratory work needed to identify viable sites. As it matures, this technology could become a significant player in the UK's energy landscape. Past estimates have suggested that deep geothermal power from the South West of England alone could meet 2% of the UK's annual electricity demand[10], potentially creating thousands of jobs in the building and running of new power plants.

Nuclear energy

- The Government will provide capital investment to establish a Nuclear Advanced Manufacturing Research Centre that combines the knowledge, practices and expertise of manufacturing companies with the capability of universities. This will complement the existing Advanced Manufacturing Centres in Sheffield and Glasgow and the Nuclear Laboratory in Sellafield. The facility will enable around 30 companies to work together on the development of processes for the manufacture of nuclear components and assemblies, to develop management

Figure 1

Financial support for technology innovation[11]

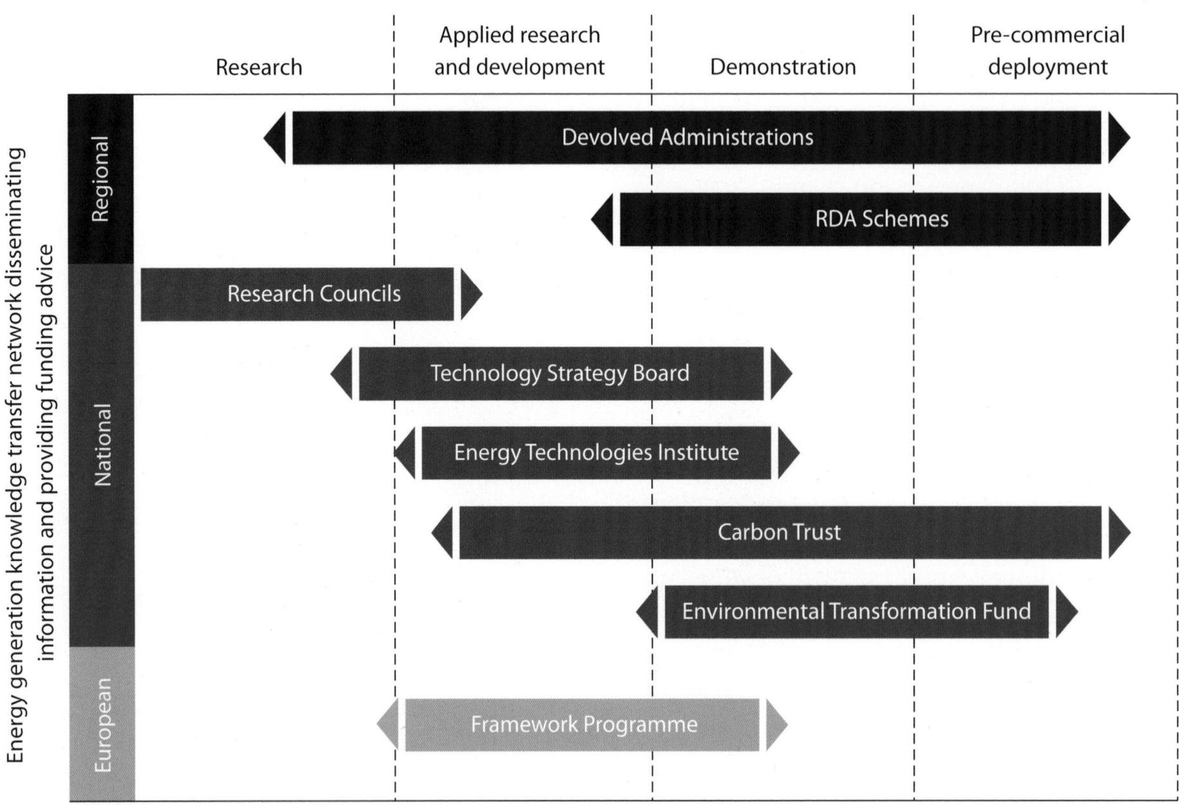

Source: Department of Energy and Climate Change

10. Camborne School of Mines (1986)
11. A range of other support for innovation is also available, for example through the Energy Research Partnership and the UK Innovation Fund

processes, training and work force development programmes and to achieve civil nuclear standards and accreditation.

Low carbon manufacturing

- *A £4 million expansion of the Manufacturing Advisory Service*, to provide more specialist advice to manufacturers on competing for low carbon opportunities, including support for suppliers for the civil nuclear industry.

In addition to the specific support provided through this new funding, the Government recognises that getting new technology off the drawing board and into production can take more than a decade and is often prohibitively expensive. Business can also face challenges in securing the right funding, support and access to facilities to test and demonstrate new technologies, and can be deterred by the high level of risk investing in new technology represents. The Government will therefore ensure professional and financial assistance is given to businesses developing new low carbon technologies (see figure 1), which shows just some of the

Box 6
Providing the support for innovation that UK business needs

- Developing a consensus on the technology that a decarbonised society might need in 2050 is essential. The Energy Research Partnership[12] will be carrying out work to focus on key research, development and demonstration milestones.
- The Government will develop a model for selecting 'technology families' with significant potential for carbon abatement and UK economic benefit, recognising that the UK's geography, capability, skills and manufacturing base gives us a strong starting point as a leader in a number of low carbon technologies.
- Developing Technology Action Plans jointly with industry, identifying the steps needed to take the technology to market. Pilots are planned for marine energy, and hydrogen and fuel cell technologies.
- The Government will ensure funding bodies become more integrated in their approach to funding technology families.
- Regional Development Agencies will play a crucial role in partnership with the Government to promote the development of market-led technology clusters for low carbon energy developers.
- The Government and the Technology Strategy Board will work to improve collaboration and knowledge sharing within and beyond the UK through the launch of the Energy Knowledge Transfer Network as a one stop shop for investors and developers in energy generation.
- The Intellectual Property Office will consider how the Government can support small- and medium-sized businesses developing low carbon technologies to license them in developing countries.

12. The Energy Research Partnership is a high level forum designed to give strategic direction to UK energy research and innovation activities.

support which Government is providing for innovation.

Alongside this action and investment, the Government wants to respond to calls from business for ever more effective delivery of innovation support for low carbon and, in particular, energy generation technologies. As a result, the Government plans to make a number of changes to the way it supports innovation, which are set out in box 6 on previous page.

The UK Innovation Investment Fund announced in *Building Britain's Future*, was set up to invest in technology-based businesses with high growth potential, including low carbon. The Government will invest £150 million to leverage equal private sector investment, which the Government believes could build this into a fund of up to £1 billion over the next ten years.

Targeted action in England's regions

The Regional Development Agencies (RDAs) are acting to provide regional leadership on energy and climate change, through the development and delivery of Regional Strategies, including energy and carbon reduction plans and targets, and supporting regional partnerships. RDAs play a key

Box 7
Skills academy ensures nuclear businesses get the workers they need

North West England is home to one of the world's largest concentration of nuclear facilities and a workforce with an internationally renowned skills base and technological expertise. The sector employs over 40,000 people and the National Skills Academy for Nuclear which is an employer led body develops skills training to support a sustainable future for the UK nuclear industry. The North West Development Agency contributed £6 million to ENERGUS in Workington. As the new flagship facility for the National Skills Academy for Nuclear, ENERGUS is a new world-class centre for high-level vocational skills. ENERGUS will directly train 250 apprentices and assist in educating under- and postgraduates elsewhere.

David Barber, head of training for British Energy, which is part of EDF Energy said: "With the increase in demand for quality skills across all sectors of the nuclear industry it is absolutely essential we have the confidence in our capability to meet this, both for our own workforce and that of our supply chain. We see the National Skills Academy for Nuclear as the key enabler to broker the provision of both provider capacity and quality skills."

Ed Miliband, Secretary of State for Energy and Climate Change, visits ENERGUS

strategic role in the regions, understanding and supporting the growth of sectors which will underpin low carbon growth such as construction, transport and low carbon technologies, and leading regionally specific activity on low carbon innovation, research and development.

To help take advantage of the unique character of each region, the Government is developing Low Carbon Economic Areas to accelerate low carbon economic activity in areas where Britain's existing geographic and industrial assets give a region clear strengths. The first Low Carbon Economic Area will be located in the South West of England and will focus on the development of marine energy demonstration, servicing and manufacture. The South West has an obvious marine resource, successful existing activity with high potential and a high level of regional expertise in marine research, development and engineering.

Supporting businesses through the global financial crisis

Swift action to minimise the risks facing the economy at home and abroad has been taken by the Government. It has been devised to ensure that the UK is best placed to make the most of the recovery when it comes about. Measures to support businesses' access to finance include: the Enterprise Finance Guarantee, Capital for Enterprise Fund and a Trade Credit Insurance Top-up Scheme to help maintain levels of trade credit insurance to ensure businesses can continue to access the finance they need.

In addition UK renewable and other energy projects stand to benefit from up to £4 billion of new capital from the European Investment Bank. The Government believes that this can bring forward £1 billion of consented small and medium-sized UK renewables projects to deployment.

Equipping our workforce with the right skills is an essential part of the transition to low carbon

Developing new skills for a low carbon economy

The growing low carbon industry in the UK can flourish only if workers have the right skills to meet the demands that businesses

Low carbon skills will be a part of every sector of the economy

will face. For example, workers in the construction sector will need the right skills to build and install small-scale renewable energy technologies, and to install the full range of measures that will make homes and businesses more energy efficient in both new and existing buildings. They will need to know how to build new low carbon infrastructure such as that required to make renewable energy and nuclear power.

Many of the skills needed are not new. The UK will need to increase the supply of science, technology, engineering and mathematics skills, for example, as well as identifying mechanisms for transferring them to new contexts. And skills already developed within the offshore oil industry, which is supported by its own skills academy, OPITO, will be highly relevant to carbon storage projects and some renewables projects offshore. But many specialist skills will be new, and a major cross sector effort will be needed to gear up the skills system to deliver those.

Key areas of activity are:

- Ensuring degree courses reflect the specialist skills essential for working in the nuclear and construction sectors. The Office for Nuclear Development is monitoring this and working alongside the sector bodies to develop a skills plan for targeted education and training.
- Renewable energy also requires specialist skills. **The Office for Renewable Energy Deployment is working with industry on a strategy for skills in wind, wave and tidal energy and is also establishing the National Skills Academy for Power**.
- Low carbon sector skills will feature prominently in a long-term active skills strategy that will be published later in 2009.

Managing the costs of transition to low carbon

There will be costs as well as opportunities from the transition to a low carbon economy. The measures set out in the Transition Plan will put upward pressure on energy prices and bills. The Government estimates that the additional impact in 2020 of the policies in this Transition Plan, relative today, is equivalent to approximately a 15% increase from current energy bills for an average business. When previously announced climate policies are included this figure is 17%.

The increase in energy bills is caused by the use of renewable generating technologies that are most expensive than electricity from fossil fuels at today's prices. However, much of this increase is offset by energy efficiency policies in this Transition Plan.[13]

Government is working to ensure that impacts are minimized for all types of business by:

- Ensuring competitive energy markets deliver affordable energy
- Incentivising energy saving
- Ensuring EU frameworks are fair to business

The Regional Development Agencies will have a key role to play in ensuring their regions maximise the opportunities emerging from the transition to a low carbon economy.

13. The 17% increase includes previously announced climate policies still having an impact on bills such as the RO and CCL. Calculations assume $80 / barrel oil price in 2020. This is dealt with in the Analytical Annex.

Chapter 6

Transforming transport

Summary

Our domestic transport currently contributes a fifth of total UK greenhouse gas emissions, and these are growing. International travel also produces greenhouse gas emissions.

This Transition Plan, along with wider policies, will cut emissions from domestic transport by 14% on 2008 levels, as set out in more detail in *Low Carbon Transport: A Greener Future*, published alongside. This chapter sets out the actions the Government is taking to encourage the development of radically different new technologies and fuels to decarbonise transport in the long-term, and to secure the oil supplies needed during the transition.

Highlights of the plan include:

- Continuing to improve the fuel efficiency of new conventional vehicles:
  - Cutting average carbon dioxide emissions from new cars across the European Union to 130g/km from 2012 with full compliance by 2015, and to 95g/km by 2020, a 40% reduction from 2007 levels.
  - To ensure that the Government leads by example, it has set targets for government departments and their agencies to procure new cars for administrative purposes that meet the EU standard for 2015 in 2011, four years early. The Government will set revised emissions requirements for new administrative cars later this year, to ensure the use and development of ever greener vehicles in its fleets.
  - Ensuring that an ambitious and realistic framework for long-term emissions

Chart 1

The UK domestic transport sector will contribute about 19% of emission savings in 2018-22

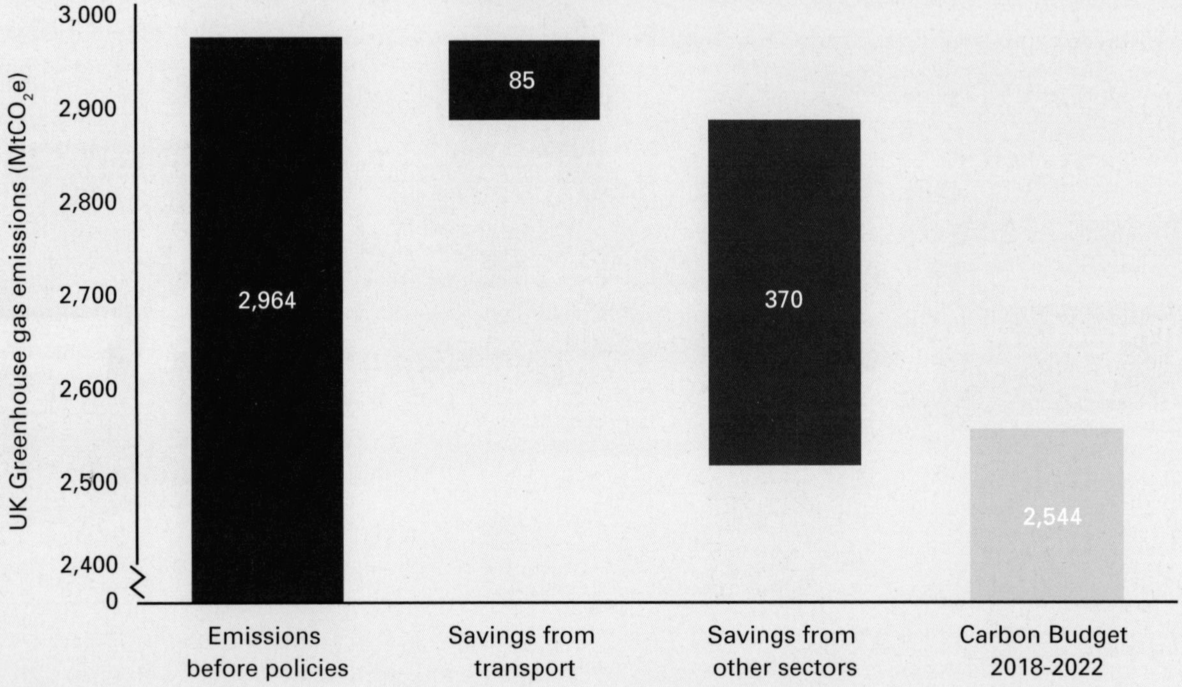

Note: Reductions due to policies introduced prior to the Energy White Paper 2007 are not shown. Savings also include interaction effects. Refer to Table A1 for a full breakdown of carbon savings in the third budget period.

Source: Department of Energy and Climate Change (2009) and Department for Transport (2009)

reductions is applied to vans through actively engaging with the European Commission as it develops proposals.

  - Investing up to £30 million over the next two years in low carbon bus technology, to deliver several hundred low carbon buses.
- Supporting the low carbon vehicles and fuels of the future:
  - Over the next 18 months around 500 electric and lower carbon cars and vans will take to UK roads through Government programmes. This includes the largest project of its kind in the world, demonstrating electric cars in real-world situations. In several cases this represents the first time that these vehicles will have been used.
  - Providing help worth about £2,000 to £5,000 towards reducing the price of low carbon cars from 2011, and up to £30 million to support the installation of electric vehicle charging infrastructure in six or so cities across the UK.
  - Committing to source 10% of UK transport energy from sustainable renewable sources by 2020.
  - Delivering a lower carbon rail system through energy efficiency improvements and greater electrification.
- Helping people to make low carbon travel decisions:
  - Providing up to £29 million in a competition for the country's first Sustainable Travel City: an opportunity for a major urban area to demonstrate how to cut car travel and increase walking, cycling and public transport use.
  - Investing £140 million between 2008-11 in Cycling England's programmes to promote cycling; plus £5 million to improve cycle storage at rail stations.
  - Funding rail and bus transport, including the England-wide mandatory bus concession offering free travel for older and disabled people.
- Requiring international aviation and shipping to reduce emissions:
  - Setting a target to reduce UK aviation carbon dioxide emissions to below 2005 levels by 2050, despite forecast growth in passenger demand.
  - Pushing hard to get a global agreement to reducing emissions from international aviation and shipping.
  - Including all flights arriving at or departing from an EU airport within the EU Emissions Trading System from 2012.
- Securing the oil supplies the UK needs during the transition, by sustaining investment in the North Sea, working to improve the functioning of international oil markets, and working with the downstream oil industry to address the issues which that sector faces.

The Government is supporting new clean vehicle technologies

The Government is helping people to make low carbon travel decisions

The scale of the task

The transport system supports economic prosperity – it provides us with access to goods and services by allowing businesses to transport these efficiently from suppliers to markets, and it provides people with the freedom to get around. Our domestic transport currently contributes 22% of total UK greenhouse gas emissions.[1] The majority of this is produced by our cars. Transport emissions from journeys made in the UK have increased by 12% since 1990. We must find cleaner ways to travel and transport goods, not only in the UK but also internationally, and secure the supplies of fuel that are required for transport.

Chart 2
Transport accounts for 22% of UK total greenhouse gases[2]

Source: Estimated emissions of greenhouse gases by National Communication source, 2007

The plan to 2020

The policies set out in this Transition Plan, and in more detail in the *Low Carbon Transport: A Greener Future*, will ensure that by 2020 we cut emissions from transport by 14% on 2008 levels and secure the oil supplies the UK needs during its transition to a low carbon economy. The Government needs to ensure that the transport system also addresses goals beyond carbon reduction, such as reducing congestion, improving safety, security and health and promoting greater equality of opportunity for everyone. The Plan will bring wider benefits, for example, local air quality will be improved and noise nuisance will be reduced with the greater use of low emission vehicles: particularly electric vehicles, and through a switch to low carbon modes of transport like walking and cycling.

There are considerable challenges to overcome:

- Our existing vehicles, fuels and infrastructure are very well established and our economy and lifestyle have built up around them. There are strong links between transport and people's lifestyle choices. Many people see little reason to make greener travel choices. Others may wish to do so, but may not have or be aware of lower carbon ways to travel.
- The transport industry has taken huge strides over the years in improving services and technology, for example safety technology. We also need to ensure that the industry directs its innovative energies towards a long-term climate change agenda. New technology requires major investment but is essential to prevent the impacts of significant climate change.

1. Results presented here may differ marginally to those in Low Carbon Transport: A Greener Future because they are generated by two different models, the Energy Model and the National Transport Model. For more detail of the differences between the models see: http://www.dft.gov.uk/pgr/economics/ntm/roadtransportforcasts08/rtf08.pdf. The domestic transport sector includes all journeys (either passenger or freight) by road, rail, air and waterways within the UK. Flights and journeys by sea that begin in the UK but end in a foreign country (and vice versa) are classed as "International Aviation" and "International Shipping" and are not counted in our carbon budgets and emissions reductions targets for the time being, due to the lack of a globally agreed methodology to allocate responsibility for these journeys to individual countries.
2. The emission and emission saving estimates in this chapter refer to greenhouse gas emissions from use of primary fuel used by road transport, non-electric trains, military aviation and shipping, domestic shipping and domestic aviation up to 2012 (after which it is covered by the EU Emissions Trading System). Emissions associated with electrified transport are included in emission figures from power and heavy industry

- The UK cannot effectively reduce international emissions in isolation. Commitment and collaboration therefore needs to be built at the European and global levels to ensure an effective approach.
- The UK is becoming more dependent on imported crude oil and is increasingly dependent on imports of diesel and aviation fuel.

The reality is that not all of the necessary changes will happen without the Government taking a strategic role, because of the nature and extent of the challenges described above. The Government's *Low Carbon Transport: A Greener Future*, published alongside this Transition Plan, sets out in more detail the action that the Government is taking now to decarbonise our transport system.[4] This Transition Plan summarises some of that action:

- The first step is to **improve the efficiency of our conventional vehicles** so they emit less greenhouse gases.
- We must move away from petrol and diesel in the long-term. So the Government is **supporting the vehicles and fuels of the future** and the radically different technologies needed.
- Cutting transport emissions is not just about changing technologies. So the Government is **helping people to make low carbon travel decisions**.
- Emissions from international flights and ships are growing and the only effective way to tackle this is internationally. So the Government is pushing hard for an international agreement to reduce **emissions from international aviation and shipping**.

Chart 3

New car and van CO_2 standards and biofuels will deliver over two-thirds of emissions savings from transport

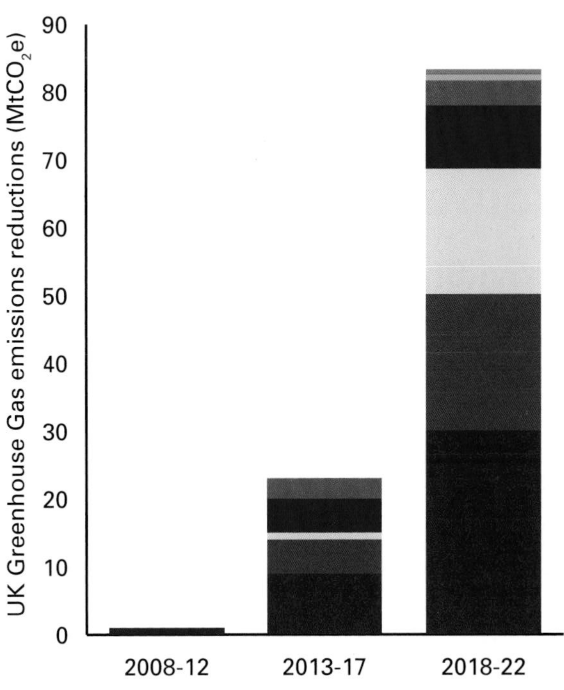

Note: Reductions due to policies introduced prior to the Energy White Paper (2007) are not shown.

Source: Department for Transport

- The Government is also acting to **secure the oil supplies** the UK needs at a fair price during the transition to low carbon transport.

4. Department for Transport (2009) Low Carbon Transport: A Greener Future

Putting the plan into practice

Improving the efficiency of our 'conventional' vehicles

Providing the right regulatory environment

In a new car today people can travel 9% further using one litre of fuel than was possible in a car manufactured ten years ago.[5]

Recognising that more could be achieved, in 2008 the EU established a mandatory target for manufacturers. From 2012, the target for average emissions from new cars sold in Europe will be 130g of carbon dioxide per kilometre, phased in to ensure full compliance by 2015. **From 2020, average carbon dioxide emissions from new cars sold in the EU must further reduce to 95g of carbon dioxide per kilometre travelled**. This represents a 40% reduction in emissions per kilometre from 2007 levels. The UK was among the leading European countries in calling for this longer-term target.

To ensure that the Government leads by example, it has set targets for government departments and their agencies to procure new cars for administrative purposes that meet the EU standard for 2015 in 2011, four years early. The Government will set revised emissions requirements for new administrative cars later this year, to ensure the use and development of ever greener vehicles in our fleets.

Chart 4

By 2020, average emissions from new cars sold in Europe will be 40% less than in 2007

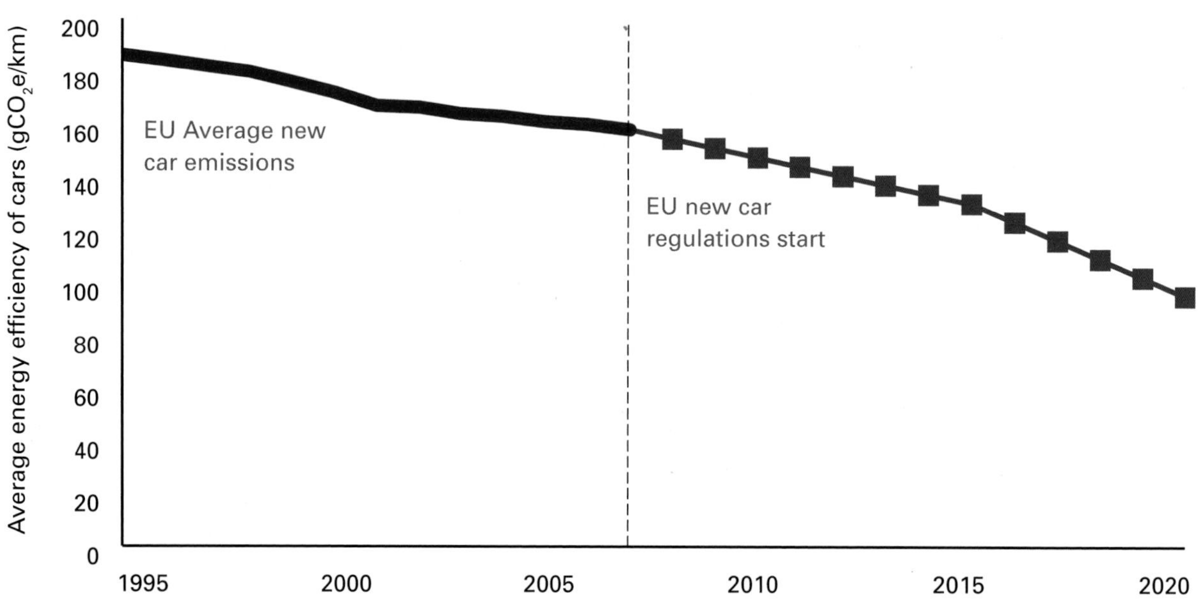

Source: Department for Transport

5. Average new car emissions today are 16.8% lower than in 1997. Source: The Society of Motor Manufacturers Traders: New Car CO2 report 2009: driving down emissions

Conventional vans can also be made to be more efficient. **The Government is determined to ensure an ambitious and achievable framework for long-term emissions reductions is applied to vans**, and is actively engaging with the European Commission as it develops its proposals.

There is big potential to reduce emissions from vans

Encouraging the cleanest vehicles onto our roads

If everyone buying a car today were to choose the most fuel-efficient model available, in the class of car they wanted, we would make huge emissions savings. To promote this, car showrooms are required to display information about carbon dioxide emissions on all new cars; this is done using a colour-coded label (see figure 1). The Low Carbon Vehicle Partnership is seeking to develop a similar label for used cars; and for vans an online database was recently launched by the Government.[6]

Figure 1

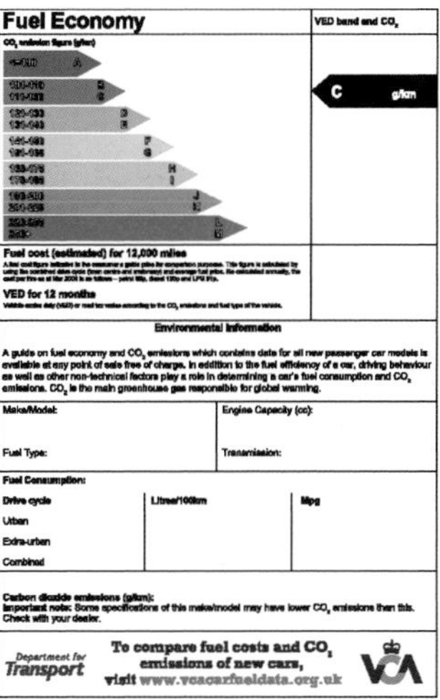

Showroom labels help car buyers save money and lower emissions by picking efficient new cars

The Government also encourages the right choices with financial incentives. For example, vehicle excise duty and company car tax have already been restructured to encourage people to choose lower emission cars.

Encouraging the cleanest buses and trains

Like cars and vans, buses and trains can be made more efficient. Hybrid buses are already on the market, and can reduce carbon emissions by 30-40% compared to conventional buses. Transport for London has 56 hybrid buses in operation, and plans to have 300 more on London's roads by 2011.

The Government will invest up to £30 million over 2009-10 and 2010-11 in low carbon bus technology, helping operators to cover the additional upfront cost of buying low carbon buses. This is expected to lead to

6. http://www.businesslink.gov.uk/vanfueldata.

The Government is investing up to £30 million by 2011 to help bus operators buy low carbon buses

the delivery of several hundred low carbon buses over the next two years. This builds on the Government's decision in 2008 to give operators running low carbon emission buses an additional 6p per km, as part of the Bus Services Operator Grant.[7]

Rail technology can also be made more efficient and is receiving similar government backing. The Government has committed to setting an environmental target for the rail industry for the period 2014-2019 (through the 'High Level Output Specification').

Box 1
Types of ultra-low carbon vehicle technology

Hybrid Vehicle: combines electric power from an on-board battery with a standard internal combustion engine running on petrol, diesel or biofuels. In a "plug-in hybrid" a larger battery is used and charged from an external source, allowing the vehicle to be used in electric mode for longer distances.

Electric Vehicle: uses a battery large enough to make all trips in electric only mode and does not use an internal combustion engine or liquid fuel.

Hydrogen and Fuel Cells: hydrogen can be used to provide energy for transport, through use in a fuel cell vehicle or internal combustion engine. In one type of a fuel cell vehicle, the hydrogen is split, and combined with oxygen from the air, which produces electricity to power the vehicle.

Supporting the vehicles and fuels of the future

In the long-term, reductions in emissions will require a radical transformation in the way vehicles are built and powered – whether hybrid, electric vehicles, biofuels or hydrogen fuel cell technology. The chart below is only illustrative, but gives a sense of the technological breakthroughs that might take us from our current conventional road vehicles, through to ultra-low carbon vehicles in 2050. The Government wants to make the UK a leading place in the world to develop, demonstrate and manufacture low carbon vehicles (see chapter 5) and has committed around £400 million to encourage development and uptake of ultra-low carbon vehicles. This includes funding for RD&D under the Low Carbon Vehicle Innovation Platform.

This timeline does not mean we have to wait for decades to see how these vehicles might look and feel to drive. Over the next 18 months around 500 electric and lower carbon cars and vans will take to UK roads

7. Low-carbon emissions buses are those emitting at least 30% less greenhouse gas emissions than a similar sized 'Euro III' bus.

Figure 2
The roadmap below shows the UK industry's view on how automotive technology will develop out to 2050

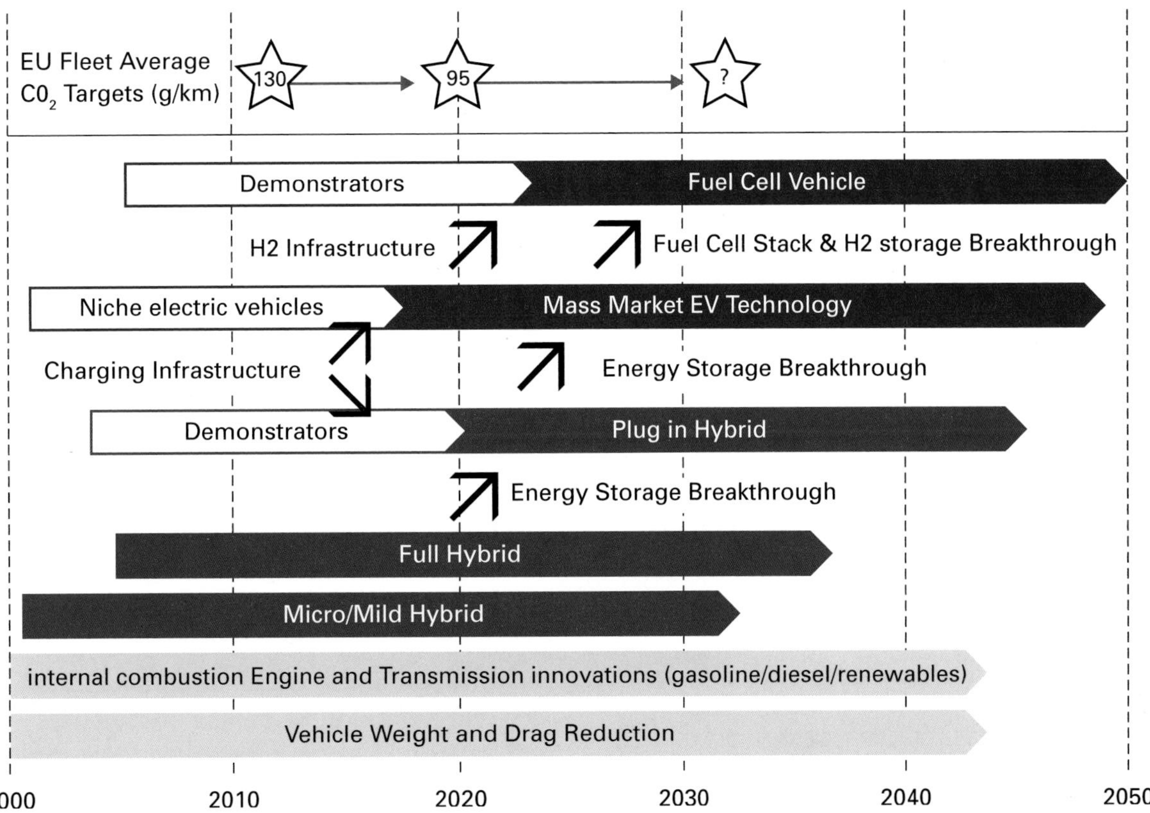

Source: An Independent Report on the Future of the Automotive Industry in the UK, New Automotive Innovation and Growth Team (NAIGT) (2009)

through Government programmes. This includes the largest project of its kind in the world, demonstrating electric cars in real-world situations. In several cases this represents the first time that these vehicles will have been used. Eight different locations will showcase the electric car trials, including Oxford, London, Glasgow, Birmingham and the North-East. In addition, 12 public sector organisations will trial low-carbon vans.[8]

To help new electric cars become competitive, **from 2011, the Government will provide financial assistance worth in the region of £2,000 to £5,000 to reduce the price of electric or plug-in hybrid cars**. This Government policy, announced in April 2009, will help make these models more competitive for motorists.

8. Programmes run by the Technology Strategy Board and the DfT's Low Carbon Vehicle Public Procurement Programme.

Moving to low carbon sources of energy in transport

As well as transforming the types of road vehicle we use, we will also need alternative fuels to replace diesel and petrol. New fuelling infrastructure may be needed to replace our existing network of petrol stations. Electricity, hydrogen and biofuels may all play a part. Diversifying our fuel sources will also help improve our energy security.

Using electricity in transport

Electricity is likely to become a major transport fuel. Current electric road vehicles are more efficient than internal combustion engine vehicles. To help kick-start the new electric charging infrastructure, **the Government is providing up to £30 million for electric vehicle charging points in six or so cities and regions, from next year**.

There is a good case for rail electrification. About 40% of the rail network is electrified accounting for about 60% of passenger travel. As well as reducing carbon dioxide emissions, electric trains are faster, more reliable, can take more passengers, and have less impact on air quality than diesel trains. The Government has undertaken work to look at this case more closely and **will shortly set out our plans for a major programme of rail electrification**.

The Government is offering up to £30 million to help install electric vehicle charging points

Further, as the UK electricity supply is increasingly decarbonised (see chapter 3) the benefits of using electricity in transport will become greater.

Using sustainable biofuels in transport

Fuel made from renewable resources such as plants or waste cooking oil, called biofuel, is already an option to complement petrol and diesel. Biofuels can play a valuable role in reducing greenhouse gas emissions and promoting security of energy supplies, by reducing our need for imported oil. The Gallagher Review warned that unsustainably produced biofuels have the potential to increase net greenhouse gas emissions, and there is a risk that certain biofuels may be displacing existing agricultural production onto areas of high biodiversity and indirectly causing greenhouse gas emissions.[9] Similarly the Government does not want to support biofuels that excessively compete for land with existing food crops, as this may contribute to food price rises.

The UK is a world leader in trying to ensure that biofuels are produced in a sustainable way, and has an ambitious biofuels research and development strategy.[10] The Government is also working internationally to establish voluntary global sustainability criteria.

Today's cars can be run on conventional fuel mixed with biofuel. The Government has made the use of biofuels mandatory through the Renewable Transport Fuel Obligation. **Suppliers must ensure that the share of**

9. http://www.dft.gov.uk/rfa/_db/_documents/Report_of_the_Gallagher_review.pdf
10. This includes £20 million investment to launch the Sustainable Bioenergy Centre; £6 million for the Advanced Bioenergy Directed Research Accelerator investigating the potential of algae for biofuels; and an intention to provide financial support for the creation by industry of a biofuels demonstration plant, which would use organic waste material to produce bioethanol and renewable power.

Our cars can already run on a mix of biofuel and conventional fuel. By 2013-14, biofuels will constitute at least 5% of the fuel blend in the UK

biofuels increases to 5% of the fuel blend by 2013-14. The Government has committed, under the EU Renewable Energy and Fuel Quality Directives, to go further to:

- **source 10% of the UK's transport energy from renewable sources by 2020**[11]
- **achieve a 6% reduction in greenhouse gas emissions from transport fuels by 2020.**

These targets will primarily be met through the use of sustainably produced biofuels.

The Government is also examining the use of biomass sourced fuels in aviation. Trials have recently been carried out by the industry. The UK is working to resolve safety and sustainability issues and the Government is encouraging the industry to plan for greater use of alternative fuels.

Helping people to make low carbon decisions

Technology is clearly important in reducing transport emissions, but other changes will also be needed. There are already choices that all of us can make, as individuals and businesses, which would reduce the environmental impact of our journeys.

Helping people to make smarter journeys

Raising awareness of alternatives to car travel can cut emissions and reduce the UK's oil needs. Large urban areas are now competing to become **England's first Sustainable Travel City. The Government announced in May that the winning area will get up to £29 million to invest over three years**. This scheme was announced following successful five-year pilots of Sustainable Travel Demonstration Towns in Darlington, Peterborough and Worcester. The Demonstration Towns reported that car trips fell by up to 9%, walking increased by up to 14%, and cycling increased by up to 12%. The towns used personalised travel planning, travel information and marketing, and improved infrastructure.

£140 million is being spent between 2008-11 to promote cycling in England. The Government is now developing a National Cycle Plan to further promote cycling; and is making available £5 million over two years to radically improve cycling storage facilities at up to ten major railway stations nationwide; and to help people to integrate rail and cycling.

11. By the end of 2014 the European Commission will undertake a review of, amongst others, the cost-efficiency of the measures to be implemented to achieve the target and the feasibility of meeting the target sustainably.

Over £10 billion will be invested in enhancing rail capacity between 2009 and 2014 to encourage people to use the railways

Rail passenger numbers have grown by 50% in the last 10 years, with further growth predicted. Over £10 billion will be invested in enhancing rail capacity between 2009 and 2014, with **overall Government support for the railways totalling £15 billion**. The case for more high-speed rail services is being explored. And **£2.5 billion per year is invested by the Government on bus services in England**, including £1 billion on concessionary fares for older people and those with disabilities.

The Sustainable Distribution Fund has helped to remove 880,000 lorry journeys from British roads with a switch to rail

Smarter distribution

To shift freight off the road and on to other modes of transport, the Government's Sustainable Distribution Fund provides grants to use rail and waterways instead. To date, this programme is estimated to have removed 880,000 lorry journeys from British roads. With funding from the Government, the Tesco train link from Daventry to Grangemouth has saved over three million miles of road journeys per year.[12]

Planning future transport networks to reduce emissions

The existing planning and appraisal processes for transport infrastructure and policy already take account of environmental impacts. Carbon dioxide reduction must now also become a greater consideration. The preparation of new Local Transport Plans and integrated Regional Strategies over the next two years represents an important opportunity to deliver this change. The Government is working with regional and local partners on best practice in transport delivery, and a suite of guidance will be available to local authorities as they develop new Local Transport Plans by April 2011.

Helping people to drive in the most efficient way

Eco-driving techniques, such as pumping up tyres and driving at an appropriate speed, reduce the engine's workload, meaning less fuel is burned and less carbon dioxide is produced. **The Government is promoting these techniques through the Act On CO_2 campaign and the Energy Saving Trust's programme of smarter driving lessons, and making eco-driving part of the driving test.**[13] Drivers can reduce their fuel use by as much as 15% after one lesson. Van and lorry drivers also receive training in these

12. The Tesco scheme received funding from the Department for Transport's Rail Environmental Benefit Procurement Scheme.
13. http://campaigns.direct.gov.uk/actonco2/home/on-the-move/driving-your-car.html

Box 2
The Government Car and Despatch Agency is leading the way with low carbon vehicles and reduced running costs

Reducing emissions and running costs is not just about buying lower carbon vehicles, but also the way that vehicles are driven. This important factor was recognised by the Government Car and Despatch Agency. The Energy Saving Trust acknowledged the Agency's driver training policy with its Fleet Heroes Smarter Driving Award in 2008. The training programme has dramatically cut fuel consumption.

Part of the approach is to monitor individual driver's fuel consumption and offer re-training where required. In addition, Green Cars, the Agency's taxi service for the Government and public sector clients, uses hybrid fuel cars. It produces half the carbon dioxide emissions of traditional black cabs. The Agency now runs the biggest low carbon taxi fleet in London.

techniques,[14] and bus drivers will be encouraged to use them through incentives in the Bus Service Operators Grant from 2010.[15]

Transport taxes also play a part: fuel duty raises revenue to help fund public services and also incentivises fuel-efficient purchases and encourages more fuel-efficient behaviour.

Reducing emissions from international aviation and shipping

Aviation

The UK's economy increasingly depends on air travel for exports, tourism and inward investment. Air travel has more than doubled since 1990 and this trend is forecast to continue over the next 20 years.[16]

Global agreement to reducing emissions from aviation is essential and the Government is pushing international forums to achieve this. The International Civil Aviation Organization has taken the first step, by agreeing global fuel-efficiency goals, discussing the need for carbon-neutral growth in the medium term, and absolute emissions reductions in the long-term. The Government will seek to build on this as part of an ambitious global deal on climate change at Copenhagen in December.

An EU government and industry body, the Advisory Council for Aeronautics Research in Europe, has set targets for aircraft manufacturers to reduce carbon dioxide emissions from new aircraft by 50% per passenger kilometre and reduce emissions of nitrogen oxides (another greenhouse gas) by 80%, relative to a 2000 base.

While action is needed at the international level, the UK will also act domestically. In some cases, additional capacity will be needed in the UK to relieve the most acute pressure points on our transport networks. In January 2009 the Government announced its policy support for the expansion of capacity at Heathrow. Alongside this the Government announced a target to reduce UK aviation carbon dioxide emissions to below 2005 levels by 2050, despite forecast growth in passenger demand. This target is the only one of its kind anywhere in the world. The Committee on Climate Change has been asked to advise on the 2050 target by December this year, including the basis for

14. Through the Government's Safe and Fuel Efficient Driving programme (SAFED).
15. From April 2010, bus operators that have improved their fuel efficiency by at least 6% over the previous two years, will receive a 3% increase in support for fuel costs.
16. Including when current economic conditions are taken into account.

the target and on the range of factors that might contribute to meeting it.[17]

These reductions are expected to be achieved from a combination of measures, including more efficient aircraft, operations and air traffic management. The Government continues to promote the use of market mechanisms such as emissions trading to incentivise improvement. It also incentivises manufacturers to develop low-carbon engines and airframes in the UK through support for priority programmes identified by the National Aerospace Technology Strategy, and tax relief for research and development investment.[18]

From 2012, all flights arriving and departing from European airports – both domestic and international – will be part of the EU Emissions Trading System.

In January 2009, the Government announced a target to reduce UK aviation carbon dioxide emissions to below 2005 levels by 2050, despite forecast growth in passenger demand

Air operators will either need to reduce their emissions through more efficient planes, demand reduction, or the purchase of emissions allowances or through auctions from other participants in the scheme.

Box 3
International aviation and shipping emissions and the UK's carbon budgets

Measuring greenhouse gas emissions from international aviation and shipping and assigning them to individual countries is complex.

The Committee on Climate Change recommended that international agreements be put in place to cover emissions from international aviation and international shipping, but concluded that they should not, for the time being, be included in the UK national carbon budgets, because of the lack of a globally-agreed way of allocating emissions to the UK, or any other country. In the case of aviation, there are also difficulties in reconciling a UK allocation system with the approach taken within the EU Emission Trading System.

However, international aviation into and out of Europe is included in the EU 2020 target to reduce emissions by at least 20% compared to 1990, with the EU Emissions Trading System providing the legally-binding mechanism to ensure aviation's contribution. The UK's carbon budgets have been set based on the EU's 2020 framework, factoring in the contribution from international aviation through the Emission Trading System. This means the Climate Change Act budgets out to 2022 and targets already take into account international aviation emissions.

17. The Committee on Climate Change is the independent body established under the Climate Change Act 2008 to advise the UK Government on setting carbon budgets, and to report to Parliament on the progress made in reducing greenhouse gas emissions.
18. Funding to support strategic R&D initiatives identified by National Aerospace Technology Strategy has been provided by the Technology Strategy Board in partnership with Regional Development Agencies, Devolved Administrations and the Engineering and Physical Sciences Research Council

The Government is also reforming air passenger duty: expanding the number of distance bands, to continue to send environmental signals to passengers and the industry alike, and ensure the sector contributes fairly to public services.

Shipping

In 2007, shipping emissions accounted for roughly 7% of greenhouse gas emissions from UK domestic and international transport.[19] However, this proportion will grow as other modes of transport decarbonise over time and with the ongoing increase in demand for global trade.

As with aviation, the Government is supporting an international policy framework to drive improvements, by:

- Working within the International Maritime Organization (IMO) to get agreement to measures to reduce CO_2 emissions from ships, when the Marine Environment Protection Committee meets in July 2009.
- Continue pressing for the agreement at Copenhagen to include a global sectoral target for maritime transport emissions that can be delivered cost-effectively and is consistent with limiting dangerous climate change to below two degrees compared to pre-industrial levels. After Copenhagen, the Government will work within the IMO to develop a new convention to deal with emissions from ships. The Government will continue to look at other options until a truly global solution can be found, or should progress within the IMO prove too slow, including proposals to include shipping emissions in a regional scheme.

Transport accounted for 75% of the UK's final consumption of oil products in 2008

Keeping our oil supplies safe and secure

We currently rely on oil for almost all of our motorised transport needs. Transport accounted for 75% of final consumption of oil products in the UK in 2008, amounting to 51.9 million tonnes of oil.

In the longer term we need to reduce our dependence on oil. As set out above, we plan to do this by improving vehicle efficiency and using new alternative fuelled vehicles. Increasing the proportion of biofuels in transport will add to the diversity and reliability of our fuel sources, although there are some risks in importing biofuels.[20]

Over the period of this Transition Plan oil will continue to be very important in meeting our energy needs, including for transport, and is

19. National Atmospheric Emissions Inventory
20. Such as crop failures or disruption in countries that produce fuels.

likely to be used longer still in aviation and shipping. Demand for oil is set to rise through to 2020 in the UK, driven by higher demand for diesel oil in motor transport and aviation fuel. The UK needs to ensure it has safe and secure supplies of the oil products it requires. The Government's approach includes:

- maximising the economic exploitation of North Sea oil and gas (see chapter 4);
- ensuring a well-functioning global oil market, and
- improving our capacity to supply fuel.

A well-functioning global oil market

The UK is becoming increasingly dependent on imports of both crude oil and key petroleum products, such as diesel and aviation fuel. In 2008, 66% of the UK's crude oil imports were sourced from Norway, with the majority of the remaining imports sourced from Russia, Algeria and Venezuela, and less than 1% from the Middle East. The UK currently relies on Middle Eastern countries, Singapore and Russia for aviation fuel. Diesel products are mainly sourced from EU countries and Russia. There is flexibility to source imports from a range of suppliers, so it is developments in the global oil market, rather than individual countries, that have the biggest impact on the security of the UK's oil supplies.

Improving the functioning and liquidity of international energy markets is important for energy security and to ensure a smooth transition to a low carbon economy. The immediate risk to oil production is not how much oil is left in the ground, but the world's ability to convert these reserves into production now and in the future. A stable and fair oil price is a key UK objective, as it preserves economic growth and ensures investment in oil production, alternative technologies and energy efficiency.

To this end, the UK has been working internationally to strengthen the dialogue between oil producers and consumers, including convening the London Energy Meeting in December 2008. The UK is continuing to pursue a range of actions internationally to try to reduce the risk of returning to excessively volatile prices, including:

- Working with international partners to **remove barriers to future investment in oil production** by encouraging stable regulatory frameworks internationally, by helping to develop the necessary skills base and encouraging further co-operation between national and international oil companies.
- **Asking the International Monetary Fund and the International Energy Agency to improve their surveillance** of what is driving oil prices and the associated macroeconomic risks of oil market developments.
- **Pushing for full commitment from partner countries to provide timely and accurate data** on oil demand, supply, stocks and investment.
- **Improving regulation, enforcement and transparency** in commodity derivative markets, through asking partner countries to implement recent regulatory recommendations from the International Organization of Securities Commissions and considering what further steps could be taken to improve information on financial flows into commodity markets.
- **Supporting the International Energy Forum's Expert Group** to develop proposals on the appropriate institutional architecture in the global oil market and consider measures to address oil price volatility.

Improving our capacity to supply fuel

The UK downstream oil industry comprises companies involved in the refining, distribution and marketing of oil products. There is a growing mismatch between UK refinery output and consumption, with increasing shortfalls in diesel and jet fuel, and surpluses of petrol and fuel oil. The UK exports surpluses to, and attracts imports from, the global oil market. The Government is working with the refining industry to review current and future challenges in refining, including the balance between product quality and product mix, future demand patterns, and the need to meet safety and environmental objectives.

Within the UK, there has been a gradual contraction in oil supply and distribution infrastructure and its capacity. Over the next 12 months, the Government will be working with industry to identify barriers to investment in new infrastructure, and develop options to strengthen future domestic supply resilience.

In relation to aviation fuel, the joint Government and industry Aviation Fuel Task Group is continuing to review demand and supply at UK airports, out to 2030, and will identify options for new fuel supply infrastructure to meet demand and improve levels of resilience by the end of 2010.

In order for the UK to be prepared for an oil emergency, the UK holds stocks of crude oil and oil products. The Government will be reviewing the industry-based compulsory oil stocks regime and, if there are better ways of delivering our international commitments to hold stocks, it will consult on options in 2010.

The Government is planning ahead to ensure our strategic transport infrastructure can cope with the climate change we expect will occur in the UK

Preparing our transport system for our changing climate

Our transport system will need to be able to cope with expected changes in climate. For example, roads and railways will need to be more resilient to heat waves and to intense rainfall, and ports and sea defences protecting coastal infrastructure will need to be more resilient to higher sea levels and storms.

The risks posed by climate change to our motorways and trunk roads are being identified by the Highways Agency in its *Climate Change Adaptation Strategy*. And Network Rail is developing 'hazard maps' to highlight rail infrastructure that may be particularly vulnerable. Through embedding adaptation into future transport planning policy, the Government is working to ensure it has the right frameworks in place to protect our strategic transport infrastructure.

Chapter 7

Transforming farming and managing our land and waste sustainably

Summary

The Government will support farmers to meet a new goal for reducing emissions while protecting our environment.

Farming and changes in land use are responsible for about 7% of UK greenhouse gas emissions. The Transition Plan, along with wider policies, will cut farming and waste emissions by 6% on 2008 levels. The long term challenge for Government and farmers is to find ways of reducing emissions while safeguarding our environment and producing food sustainably, especially as the climate changes – but the transformation can start straight away.

The Transition Plan will also help protect the equivalent of over 37 billion tonnes of carbon dioxide that is currently locked into natural reservoirs of carbon in our soils and forests – an amount that goes up or down depending on how land is used. Good land management is vital for keeping these stores absorbing and locking away carbon dioxide rather than emitting it.

When waste decomposes it releases methane, and this makes up about 4% of total UK greenhouse gas emissions. The Transition Plan will cut waste emissions by 13% on 2008 levels.

Chart 1
The farming, land and waste sectors will contribute about 4% of savings in 2018 – 2022

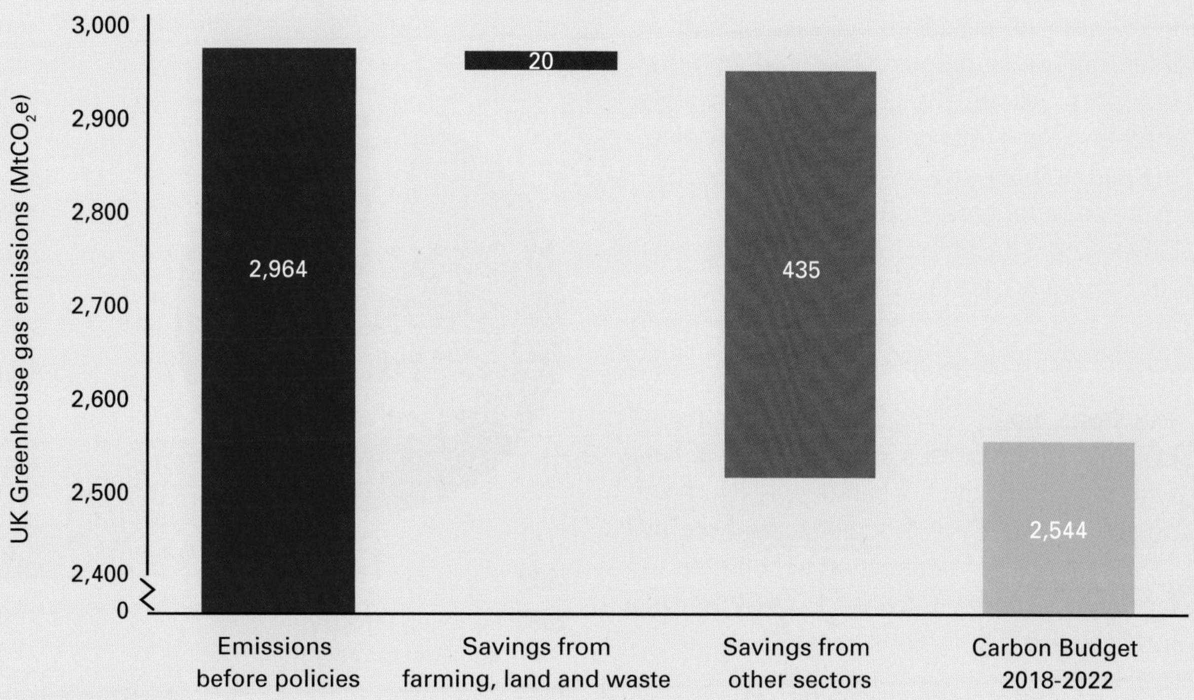

Note: Reductions due to policies introduced prior to the Energy White Paper 2007 are not shown.
Source: Department of Energy and Climate Change

Highlights of the Transition Plan include:

- Encouraging English farmers to take action themselves to reduce emissions to at least 6% lower than currently predicted by 2020, through more efficient use of fertiliser, and better management of livestock and manure.
- Reviewing voluntary progress in 2012, to decide whether further Government intervention is necessary. The Government will publish options for such intervention in Spring 2010.
- Ensuring comprehensive advice programmes are available to support farmers in achieving this aim, to reduce their emissions from energy use, and to save money in the process.
- Researching better ways of measuring, reporting and verifying agricultural emissions.
- Encouraging private funding for woodland creation to increase forest carbon uptake.
- Support for anaerobic digestion, a technology that turns waste and manure into renewable energy.
- Reducing the amount of waste sent to landfills, and better capture of landfill emissions.

The scale of the task

The vast majority of farming emissions come from methane produced by livestock and their manure, or nitrous oxide produced from fertilisers. Waste's impact comes mostly from methane from rotting rubbish. By designing the carbon budget system to take account of these other greenhouse gases, the Government has introduced the first formal UK framework for tackling the impact of these sectors on climate change (see chapter two for more on carbon budgets).[1]

This framework will help farming to remain a strong and prosperous industry while joining with the rest of the UK in reducing its impact on our climate. Such reductions are essential – without new and concerted action then farms will account for over a third of the UK's total allowable emissions by 2050.

Of these sectors, the bigger long term challenges are in farming, where:

- There are physical limits to how far emissions can be reduced. Farming involves complex natural cycles such as the gases produced by livestock reared for meat, dairy or wool.
- The world population will rise to more than nine billion by 2050 according to the United Nations. The UK must play its role in ensuring safe, affordable food supplies, balanced by the need for the sector to adapt to the impacts of climate change and safeguard environmental resources such as biodiversity and water quality.
- Agricultural products are traded internationally. So in reducing emissions in the UK we need to make sure that we do not simply transfer the problem to other countries.

Chart 2

Farming, forestry, and land management are responsible for 7% of our greenhouse gas emissions, and waste is responsible for 4%[2]

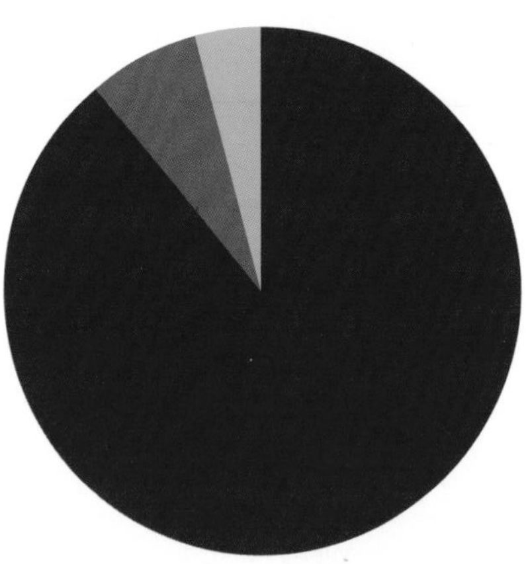

Other 89%
Farming, forestry, and land management 7%
Waste 4%

Source: Estimated emissions of Greenhouse Gases by National Communication source category and end user: 1990-2007

- Compared to other sectors, there are much larger scientific uncertainties in estimating agricultural emissions and predicting the effects of changing practices. For example, the amount of nitrous oxide released from spreading fertiliser can depend on the soil type, the weather conditions, when and how the spreading was done, and many other factors.

1. The Government decided to include greenhouse gases like methane and nitrous oxide in the UK carbon budgets in line with advice from the independent Committee on Climate Change (see Building a Low-Carbon Economy: The UK's contribution to tackling climate change pp 335 – 362, available from http://www.theccc.org.uk/reports/building-a-low-carbon-economy).
2. The emissions and emission saving estimates in this chapter refer to greenhouse gas emissions from farming (including CO_2 emissions from farm machinery other than tractors), land use, land use change and forestry, and emissions from landfill.

Farmers, land managers, foresters and other stakeholders must be involved in developing a consensus about the solutions, both at home and abroad, to these kinds of long term questions – but there are changes that scientists tell us should be made straight away to reduce emissions and bring other economic and environmental benefits.

The plan to 2020

This Transition Plan will:

- Encourage English farmers to take action themselves to reduce yearly emissions from livestock and fertiliser by the equivalent of more than three million tonnes of carbon dioxide, compared with their current projected levels.[3] By improving fertiliser efficiency, manure management, and livestock feeding and breeding, farming emissions will be 6% lower than they would otherwise be.[4]
- Cut England's yearly waste emissions by the equivalent of one million tonnes of carbon dioxide by 2020, on top of the reductions already predicted. This will reduce UK waste emissions to 13% below today's levels.

Chart 3

Farming and waste will deliver most of their savings in 2018 – 2022

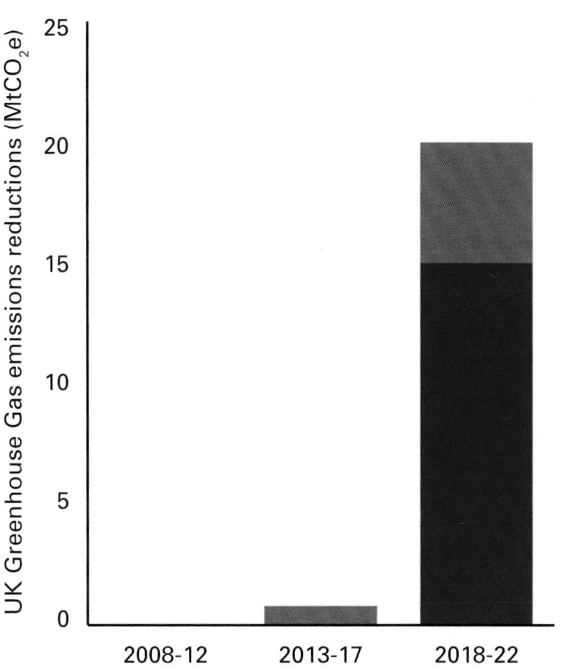

Note: Reductions due to policies introduced prior to the Energy White Paper 2007 are not shown.

Source: Department of Energy and Climate Change

This plan involves:

- transforming farming
- protecting and increasing our natural carbon stores, and
- reducing emissions from waste.

3. This figure is averaged over 2018 – 2022.
4. Both historical figures and projections for agricultural emissions come with very large uncertainty ranges. Figures in this chapter have been calculated using central estimates.

Putting the plan into practice

Transforming Farming

Setting an aim

The Government's intended saving for farming is based on the best available evidence, which shows it to be ambitious but realistic. The Government may set out a higher goal in future as evidence of effective practices grows[5], but this demonstrates the level of effort the farming sector needs to make, which must begin now and increase progressively in the years ahead.

The Government wants to see a strong and competitive farming industry that starts making real progress in tackling its emissions. Better livestock management will be an important part of this process.

Because most of the actions the Government expects farmers to take to reduce emissions also save money or increase productivity, it expects to see a proactive response from the sector and its leadership. The Government is now calling for the sector to agree on a voluntary basis, by Spring 2010, an action plan for reducing emissions.

The Government will review action taken by the sector in 2012 to decide whether voluntary action will be sufficient in getting agricultural emissions down far enough and fast enough. This review will use the best available proxy indicators of success while research to improve the measurement of agricultural emissions is still underway.

The Government will work with farmers, delivery bodies, and the Devolved Administrations to develop a shortlist of options for intervention to be triggered in case of insufficient progress.

This will consider new policies – regulatory, economic, voluntary, and advisory – including those used in other sectors of the economy and other countries. In addition to new initiatives, some current policies could be shaped to bring greater focus on climate change mitigation – for example, Environmental Stewardship, Nitrate Vulnerable Zones, and the England Catchment Sensitive Farming Delivery Initiative. The Department for Environment, Food, and Rural Affairs (Defra) will set out this shortlist as part of its climate change strategy in Spring 2010.

Focusing on cost-effective action by farmers

The Government recognises that globally, scientific evidence on how best to reduce agricultural emissions is at an early stage, and UK-focused studies commissioned by the Committee on Climate Change and by the Government do not yet agree on all the answers.

However, all the studies agree that there are opportunities to reduce emissions and save money by making agricultural processes more climate friendly – including by using fertilisers more efficiently, and by improving livestock feeding, breeding and

5. The intended saving for farming will also need to reflect the Government's improving understanding of historic and future farm emissions. If scientific progress shows that today's assumptions about current and future emissions are wrong then the intended saving may shift in line with the revised figures.

manure management. In common with other businesses, farmers and land managers must also benefit from reducing their emissions from energy use, including through energy efficiency measures that can save money.

Supporting action by farmers and land managers

The Government believes that farmers and land managers should make the decisions on the ground about how to manage their land and their businesses, and it will support them to make voluntary changes to reduce all types of greenhouse gas emissions, which must be tackled together at the farm level.

The sector has already begun to take action. Key organisations formed a joint Climate Change Task Force in January 2007,[6] and the Government also values the commitment and expert advice of the Rural Climate Change Forum[7] in helping to tackle this issue and promote a thriving farming industry that meets consumers' demands in a sustainable way.

Advice like this helps inform practical help from the Government, which includes bioenergy capital grants[8], financial support for anaerobic digestion demonstration plants, England Catchment Sensitive Farming Delivery Initiative[9], and payments under Environmental Stewardship.[10]

Spreading fertiliser on Cambridgeshire wheat: The Government will help farmers use less fertiliser to save money and reduce emissions at the same time.

To provide further support, the Government will now:

- **Provide advice, and demonstrate how to take cost-effective action.**

Box 1
Case study: efficient use of nutrients

At Thorney Abbey Farm in Nottinghamshire, Andy and Sue Guy have saved money and reduced emissions through reducing the amount of inorganic fertilisers they use, by:

- carrying out their own analysis of the nutrient content of the manure and slurry from their livestock, and using this instead of inorganic fertiliser
- choosing the crops they grow to match them with the availability of nutrients in manures and soil, and
- carefully considering when and how they apply fertiliser to get nutrients to the plants at the right time.

Andy and Sue say that "the measures we've taken already have been quick and simple, with minimal outlay, but have yielded tremendous savings".

(Source: Farming Futures, http://www.farmingfutures.org.uk/)

6. The taskforce includes the National Farmers Union, Country Land and Business Association, and the Agricultural Industries Confederation.
7. This Forum brings together the key organisations with an interest in the rural sector – see http://www.defra.gov.uk/environment/climatechange/uk/agriculture/rccf.
8. See http://www.bioenergycapitalgrants.org.uk/
9. See http://www.defra.gov.uk/farm/environment/water/csf/index.htm
10. See http://www.naturalengland.org.uk/ourwork/farming/funding/es/default.aspx

The Government is already listening to and discussing these issues with farmers through the Rural Climate Change Forum, Farming Futures project, and the 'climate-friendly farming' theme of the Act on CO_2 campaign, launched at the Royal Show in July 2009. The Government will now work with Natural England, the Environment Agency, Regional Development Agencies, the Carbon Trust, and other partner organisations to identify gaps in existing low-carbon advice, and develop more comprehensive services to fill them. This will make best use of Business Link, the Government's single portal for advice to business.

- **Support for energy efficient and low carbon farming.** Within the limits imposed by the current EU rules on state aid, the Government and the Carbon Trust will work to make farming businesses eligible for its interest-free loans for low-carbon activity (see chapter 5). The exact size of the loans available to farmers will depend on interest rates in the economy and whether they have received any other state aid, but the Government estimates that the maximum loan will be in the region of £30,000.

Figure 1

Anaerobic digestion can be used to convert different types of waste into renewable electricity, heat, and biofertiliser for farming. It can also be used to make transport fuel or gas (not shown)

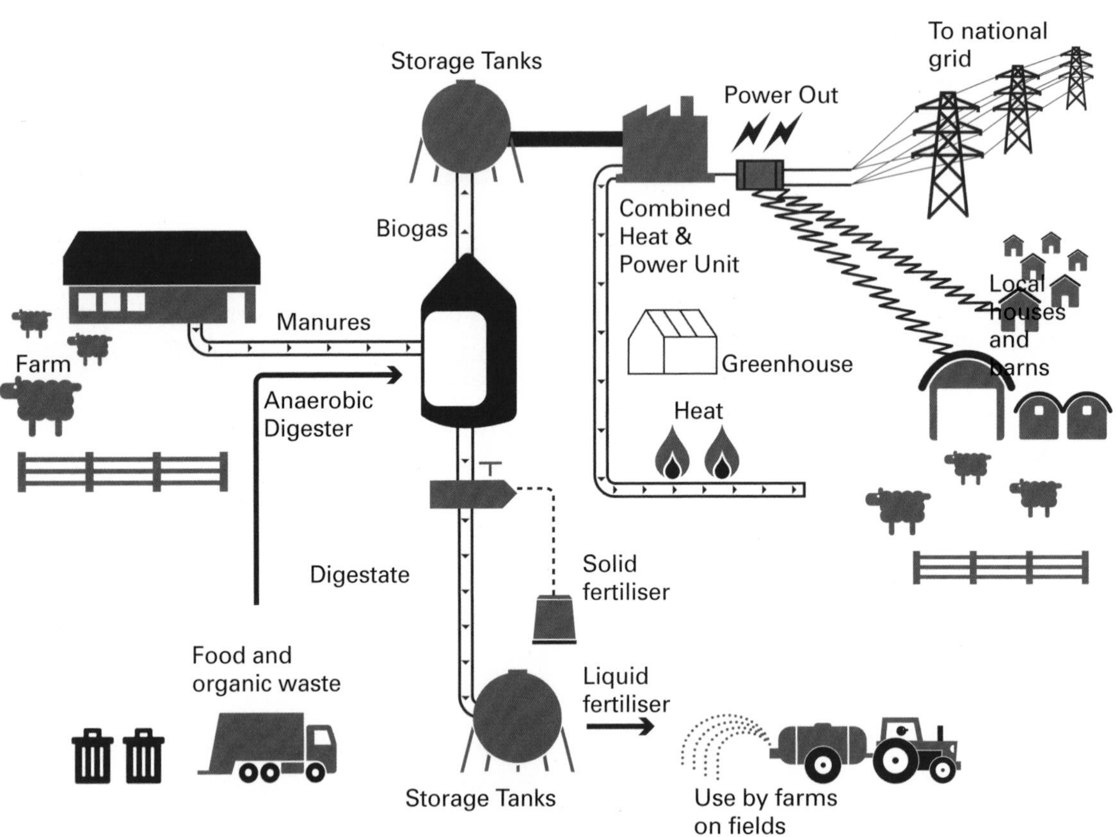

- **Improve emissions measurement**. Farmers can expect the real-world impacts of changes in their practices to be captured in national carbon accounting because of a new Government research programme. This will improve the measurement, reporting, and verification of emissions, which has been shown to be possible in other countries such as New Zealand.
- **Maximise the potential of anaerobic digestion to produce renewable energy from farm waste and reduce emissions from manure.** A Government-appointed Anaerobic Digestion Task Group has made recommendations for an Anaerobic Digestion Implementation Plan, published alongside this White Paper.[11] The Government will respond later in 2009.

Decarbonising our whole food chain

Through these steps the Government will help to make agricultural production in England as greenhouse-gas-efficient, cost-efficient, and competitive as possible. To find a long-term, sustainable way to feed ourselves safely and affordably in a low-carbon world, the Government is also looking at how emissions throughout the food chain are affected by decisions by business and consumers – including what we buy and eat, and how much we waste.

The Government is committed to reducing emissions throughout the food chain, including greenhouse gases emitted during the distribution of food

Protecting and increasing our natural carbon stores

Soils and forests are large natural reservoirs of carbon: they lock away carbon dioxide, storing it through tree growth and various natural processes in a way that avoids it contributing to climate change. Scientists estimate that UK soils and forests are currently absorbing a small amount of carbon from the atmosphere every year.

The total carbon stored in UK soils and forests is equivalent to over 37 billion tonnes of carbon dioxide – more than 50 times the UK's annual emissions. Changes to the landscape (including building work, soil tilling and forest management) need to be done in a way that protects and where possible grows these stores, particularly as climate change itself is expected to affect natural processes in a way that could cause some of this store to be lost.

We need an approach that reflects all our needs – including reducing emissions, protecting and growing carbon stores, and providing goods and services to society in a sustainable way.

11. See http://www.defra.gov.uk/environment/waste/ad/implementation-plan.htm

Looking after our soils

To fill gaps in our scientific knowledge – including whether the UK is losing or gaining soil carbon – the Government is spending £1.5 million each year on soil research. This includes work on protecting or increasing soil carbon, the likely effects of climate change (changes in temperature and moisture can lead to the breakdown of more organic matter and losses of carbon dioxide to the atmosphere), and new ideas like using "biochar" as a soil additive, which could lock up carbon dioxide in the soil and might also have useful benefits for crops.

Given the big potential impact of losing soil carbon, the Government is also taking action now where it is safe to do so, including a requirement on farmers to take steps to protect their soil. Over half of the UK's soil carbon is stored as peat, so the Government is working with industry and stakeholders to reduce the extraction of peat for uses like multi-purpose compost and 'grow' bags. Over half of this market was 'peat-free' by 2007.

Farmers' management of soil is vital for protecting this important store of carbon.

Protecting, managing, and growing our forests

In 2007, forests in England removed a net total of about 2.9 million tonnes of carbon dioxide from the atmosphere. This removal rate is declining, as forests planted in the 1950s to 1980s reach maturity. If woodland creation and removal continue at their 2007 rates, it will drop to around half a million tonnes per year by 2020, and if woodland creation stops entirely it will fall to only a hundred thousand tonnes.

Woodland creation is a very cost-effective way of fighting climate change over the long term, but it requires an upfront investment. The Government is already doing this: woodland creation represents 60% of the grant aid administered by the Forestry Commission. But to realise the potential for 2050, we need to see a big increase in woodland creation – and we need to plant sooner rather than later.

The Government will support a new drive to encourage private funding for woodland creation. If we could create an additional 10,000 hectares of woodland per year for 15 years, those growing trees could remove up to 50 million tonnes of carbon dioxide between now and 2050. Well-targeted woodland creation can also bring other benefits, including a recreational resource, employment opportunities, flood alleviation, improvements in water quality, and helping to adapt our landscapes to climate change by linking habitats to support wildlife. The Government will ensure that woodland creation policies continue to respect the benefits and demands of landscape, biodiversity and food security.

This will allow businesses and individuals to help the UK meet its carbon budgets, whilst delivering the other benefits that woodlands can bring. A number of informal schemes already exist, and the Government will work with them and with the private sector to consider how it can build on and complement existing initiatives. The Government is already laying the groundwork: including through the consultation on a Code of Good Practice for Forest Carbon Projects led by the Forestry Commission, and the Government consultation on corporate carbon reporting guidelines[12], which sets out how funding for domestic emissions reduction projects can be reported in company accounts.

The Government will encourage the creation of new woodland to lock away carbon.

Box 2
Are forests a reliable way to remove and store carbon?

Some people have raised concerns that carbon stored in forests is not safely locked away, because the trees could die or be chopped down and then release carbon as they rot. In the UK, we have a strong regulatory framework to protect woodlands and promote sustainable forest management – meaning that following felling, trees are replaced to ensure continuous woodland cover. Although the amount of carbon stored in the woodland temporarily declines following felling, this cycle reduces overall emissions in the longer term when the wood is used as a source of energy (woodfuel), or as a building material (where the carbon will stay locked up, and emissions from producing more energy-intensive building materials are avoided). The carbon stored in trees and wood products is accounted for by the system that the UK uses to record and report its greenhouse gas emissions.

The draft Code of Good Practice for Forest Carbon Projects, published by the Forestry Commission in July 2009, will increase confidence in woodland creation projects by providing guidance on how to properly account for stored carbon, and by setting out requirements which projects must meet to become accredited.[13]

12. See http://www.defra.gov.uk/corporate/consult/greenhouse-gas/draft-guidance.pdf
13. See http://www.forestry.gov.uk/carboncode

Box 3
Encouraging woodland planting – The National Forest Company

In the English Midlands, The National Forest Company has attracted funding from a wide range of private sector companies, including SMEs, who choose to have their name connected with creation of woodland in the National Forest. Their research has shown that investors are attracted by a range of benefits, including the recreational facilities and wider environmental benefits provided by woodlands. Fighting climate change may not be at the top of their list initially, but moves up the agenda as they learn through their involvement with the forest. This means that as well as the carbon storage and other benefits, the scheme is helping to raise awareness of climate change.

Reducing emissions from waste

The Government's strategy is to reduce the overall level of waste produced by households and business, and to send less waste to landfill – particularly substances that will breakdown to produce methane, a potent greenhouse gas.

Emissions from waste are already down 62% from 1990 levels, and predicted to fall further from existing policies alone. But as it degrades, biodegradable waste produces methane, a powerful greenhouse gas – often for many decades. So the Government has decided to go further by:

- Reducing the amount of waste produced. The Love Food Hate Waste campaign[14] is reducing household food waste by over 250,000 tonnes, and the Government's recently announced measures to improve food labelling will have a further impact.
- Putting even less of the waste we produce into landfills. The Government will encourage greater production of bio-energy, particularly from combustion. It also plans to encourage more processing of food waste, agricultural waste, and sewage using 'anaerobic digestion' to produce biogas. The UK Government will be consulting later this year on banning certain materials or types of waste from landfill, including the most climate damaging substances; any such bans would be expected to come into force no later than 2020 and would work alongside polices such as landfill tax.
- Capturing more of the methane produced from existing landfill. The UK Government has asked the Environment Agency to consider ways in which control over landfill gas emissions could be tightened including, if necessary, tighter regulation.

Uncertainties remain about the science of landfill gas emissions and the best measures to reduce them, but the Government estimates that extra measures of this

14. This campaign is being run by the Waste & Resources Action Programme (http://www.wrap.org.uk/) and Waste Aware Scotland (www.wasteawarescotland.org.uk).

Box 4
Protecting our marine environment

This chapter focuses on the sustainable management of our land, but the marine environment is also important. The world's oceans currently take up around one quarter of the carbon dioxide we emit, but as they store more carbon the rate at which they can absorb our emissions is slowing.

The more carbon dioxide the oceans absorb, the more acidic they become. This is a serious threat to many marine organisms and could have wider impacts on food webs and ecosystems.

The Government is funding work to improve our understanding. In 2005, it helped to found the Marine Climate Change Impacts Partnership, which aims to make evidence and advice available to policy advisors and decision-makers. Earlier this year, the Government announced a five-year £11m study[15] into ocean acidification, funded jointly with the Natural Environment Research Council.

The Marine and Coastal Access Bill, currently before Parliament, introduces a new system of strategic marine planning with sustainable development at its heart. This will enable the Government to take a more managed and coherent approach to the ways in which we use our marine resources.

The Government is working to protect the ocean as a carbon store and as a habitat for marine life © 2007, Defra, JNCC, Marine Institute, British Geological Survey, University of Plymouth.

type, together with the recently announced increases in landfill tax, might yield a further one million tonnes a year savings in carbon dioxide equivalent by 2020.

15. See http://www.defra.gov.uk/news/2009/090428.htm

Chapter 8

Developing a roadmap to 2050

Summary

The Government's policies to 2020 set us on the path to transform the UK into a low carbon economy by 2050. However, the scale of the challenge of our goals for 2050 means that the Government, Ofgem, industry and consumers need to consider now, the possible pathways to achieving this goal – through 2030 and beyond. While no-one can say with certainty what the energy system will look like in 2050, there are two things that are likely to be true:

- It will be more achievable and less costly to decarbonise the energy system if we significantly reduce overall demand for energy.
- We will need at least as much, if not more electricity, which will come from a variety of low carbon sources.

Both of these will involve substantial structural and behavioural change from all sectors of society, including large-scale public and private investment.

It is not possible to predict with complete certainty what 2050 will look like. Areas where there is greatest uncertainty include:

- How to balance the challenges associated with reducing energy demand with the costs and implications of increasing low carbon energy supplies.
- The best combination of sources of low carbon energy, and in particular what technologies we will rely on to meet our long-term heating and transport needs.

The Government will work with industry and stakeholders to articulate a roadmap to a low carbon UK in 2050

Chart 1
An illustration of how electricity demand could increase, even as overall energy use declines

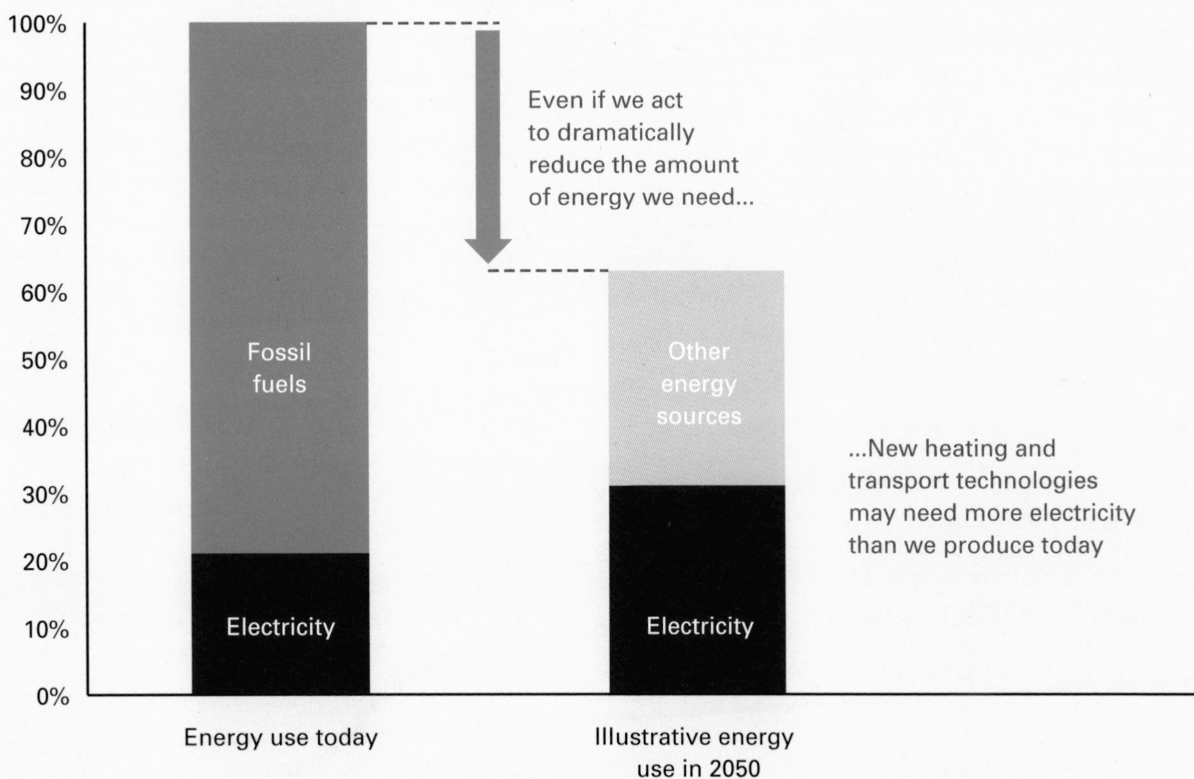

Even after we have reduced our energy demand significantly, changing technologies for transport and heating could possibly increase our use of electricity by 2050.

Source: Department of Energy and Climate Change illustrative scenario

Note: Other energy sources likely to include a mix of bioenergy, hydrogen, residual fossil fuels and other sources of primary energy.

What is clear, however, is that securing low carbon energy for the longer term is a challenge that requires investment, innovation and integration of a scale without precedent in the energy sector. Given the importance of the task and the public policy choices involved, the Government will: continue to develop a strategic framework to decarbonise the UK based around carbon pricing; enable and shape major infrastructure and capacity building investments which have long lead times; and build consensus between industry, the Government and the public on the scale and nature of the changes required. The Government will publish a roadmap setting out the path to 2050 by spring 2010.

This chapter looks at some of the changes needed as we move through 2030 and towards 2050, and explains how the Government will work with industry, stakeholders and other experts to develop a more detailed roadmap to ensure that choices made today make sense for the long-term. It is important in doing so to recognise uncertainty in looking so far into the future; by 2050 we will use a wide variety of technologies, some of which may not yet be invented. Therefore, any roadmap will need to:

- plan for action where the direction is clear
- be responsive to changing circumstances as we move forward
- foster testing and development of various options and technologies, and
- continue to build on a framework including a carbon price driving technological and behavioural change.

There are a wide range of views on the best way to achieve our emissions targets. They have some common themes and goals, but with different means of achieving them. For example, recent analysis by the Committee on Climate Change and the CBI propose different policies and paths to decarbonise our electricity generation.[1] As we move forward, the Government will continue to assess the full range of analysis as it develops and take action where appropriate to guide our path..

Planning now through 2030 to 2050

Decarbonising our energy system will be a gradual process over the coming decades (chart 2). The primary driver of this decarbonisation will remain the three broad interventions set out by Lord Stern's *Review of the Economics of Climate Change* (2006): putting a price on carbon emissions; driving new technologies; and helping people make low carbon choices. We will continue to use significant quantities of fossil fuels during the transition and may well do so even beyond 2050 (although in combination with technology that captures most of their emissions).

Chart 2

There are a range of scenarios for CO_2 emissions in 2050 that are in line with the UK's overall greenhouse gas emissions target

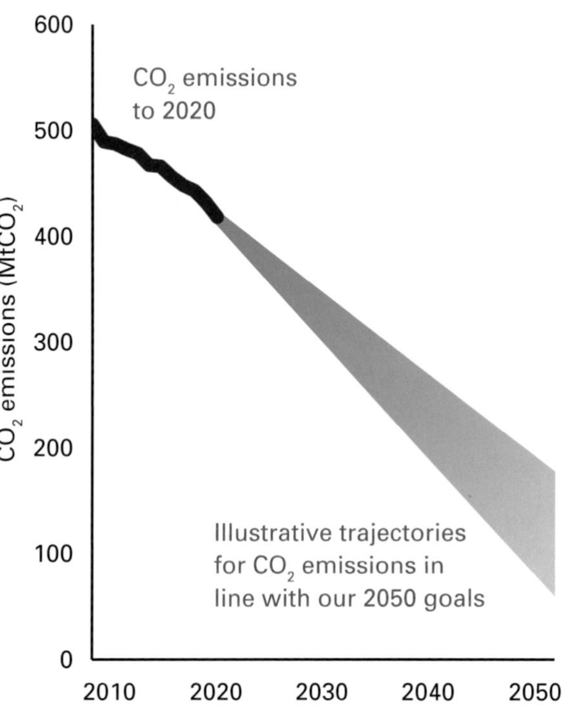

The trajectories for CO_2 emissions to 2050 depend on a number of things, including levels of non-CO_2 greenhouse gas emissions and carbon trading. In practice, the most cost-effective pathway to achieving our emissions targets could require more reductions (in absolute terms) early on than the illustrative straight-line trajectories above suggest.

Source: Department of Energy Climate Change (2009)

1. The CBI *Decision Time – Driving the UK towards a sustainable energy future* (2009) and the Committee on Climate Change *Building a Low-Carbon Economy - the UK's contribution to tackling climate change* (2008)

As the changes expected by 2050 are so large and the time it takes to achieve them is so long, some of the decisions made today will have an impact on what the energy system will look like in 2030 and beyond. These changes will have significant implications for industry, businesses and consumers, so it is important that we develop a shared understanding of the choices available, the practical challenges and constraints to be managed in implementing them and their implications for society. Although we cannot know all the details of our future energy mix, we need to examine now whether we have the necessary building blocks in place to support this longer term transition. We must also consider the nature and timing of future key decisions in view of the necessary trajectory beyond 2020.

The Government has started this process and sets out here some of the issues and the further work we will carry out, with industry and others, including:

- The initial conclusions we can draw from existing analysis.
- What the role of reducing demand for energy might be.
- What plans we need to put in place to cope with a potential increase in the demand for electricity.
- The largest areas of uncertainty that need to be managed on the path to 2050.
- The practical challenges that will need to be met.
- How we will go about developing the roadmap we need to secure long-term changes.

The overall goal will remain to meet our emissions and renewable energy targets in a way that minimises the cost to our economy and maximises our living standards.

Chart 3
Modelling suggests a range of scenarios for 2050

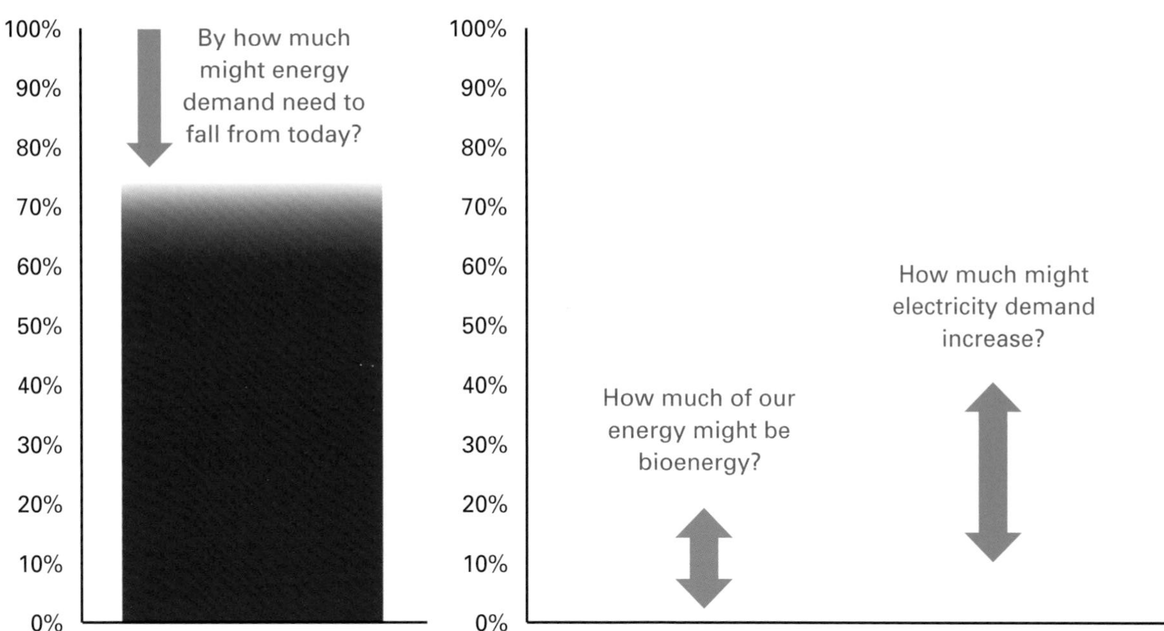

Source: Representation of model results from the MARket ALlocation (MARKAL) model. Please refer to the Analytical Annex for more detailed discussion

Initial conclusions from the analysis

A range of analysis by independent experts and by the Government has described what a low carbon UK in 2050 might look like and what we need to do to get there. How we get to 2050 depends on technology costs, physical constraints (including how fast infrastructure can be built), global energy prices, global emissions targets, decisions made by individuals and businesses and the future policy choices of the Government.

Chart 3 shows a range of outcomes from a set of scenarios describing what the energy system might look like in 2050. It shows that despite significant uncertainties, some challenges to our energy use and supply are clear enough for the Government, industry and consumers to debate, expect and plan for.

We can be certain that all sectors are likely to undergo substantial changes, although the effort and rate of change in any one sector will depend on the degree of effort and rate of change in other sectors too.

For example, although the optimal rate at which to reduce emissions from electricity generation will depend on how our heating and transport systems evolve, the Government recognises that most analysis points to the need to aggressively reduce emissions from electricity generation and will continue to act to do so.

We will also need to continue reducing our non-carbon dioxide emissions (from, among other sources, agriculture, waste and industrial processes), as well as our emissions from aviation and shipping, and progress in these sectors will have a direct impact on the scale of emissions reduction needed from heat, power and transport.

Where there are large uncertainties the Government will focus on understanding when pivotal decisions need to be taken, and what the sensible interim actions are to reduce the uncertainty and make the transition faster and more efficient.

While the analysis suggests that a wide range of scenarios and pathways could allow us to meet our goals, there are two common themes that emerge:

- Demand for energy needs to reduce dramatically through 2030 to 2050 – in some projections by over 40% from 2005 levels.
- Despite this, we can expect to see demand for electricity to at least stay the same and probably increase. (for example, as more electricity is used in the heating and transport sectors). In some scenarios, demand could increase by as much as 50% – to deliver this and meet our emissions targets the power sector must comprehensively switch to low carbon sources.

Reducing energy demand

Reducing our demand for energy from the energy system is fundamental to the Government's strategy, particularly because in many cases doing so saves money for households and businesses, whilst maintaining or improving our standards of living. Using technologies like smart meters and a more flexible grid will enable us to match new sources of low carbon energy with demand more efficiently. If we reduce demand for energy, it will be cheaper and more practical to decarbonise the energy system.

Demand for energy can be reduced by:

- Increasing the efficiency with which we use energy (for example, through insulating our homes or commercial buildings).

- Reducing our demand for energy-intensive services (for example, through turning down our thermostats, when practical).
- Making better use of our primary energy sources (for example, by using the recycled heat from power stations).
- Using other forms of primary energy (for example, heat from the ground and the sun to warm our buildings).

A dramatic reduction in demand will require considerable effort in all parts of society – the homes we live in, our workplaces, the appliances we use and the way we travel may need to be significantly different. This will need to happen in a world where we expect our population and economy to have grown, where many of us will have the ability to purchase more goods, travel more and live in bigger and better homes.

Some of the technology needed to improve energy efficiency already exists, and it is likely to improve further. The policies on energy efficiency set out in previous chapters will set us on the right pathway to 2020, but as they are implemented the Government, businesses and consumers will need to continue to look for further opportunities to reduce demand.

Therefore, there needs to be society-wide discussion about the practical implications of such demand reductions, including both the physical changes to improve energy efficiency and the changes in attitude to energy use that might be involved. The challenge of doing so needs to be evaluated against the practicality and cost of developing additional low carbon energy, to decide the appropriate balance of effort.

An increase in demand for electricity

Even as we reduce our demand for energy, we are likely to have an increased need for electricity. This is because many (but not all) scenarios suggest that in the future, more of the energy for heating and travel could come from electricity.

Power stations in general have long life spans (for example, a coal plant can continue to be used for well over 30 years, and a future nuclear power plant could stay online for up to 60 years or more). Therefore, given the need to replace significant amounts of generating capacity in the next decade (see chapter 3), investment decisions being made now will shape our generating mix out to 2050 and beyond.

We therefore need to plan for this today and as we do so, we need to consider both the generation mix and the generating capacity that may be required.

The supply of electricity will need to be almost entirely decarbonised. In order to do this it is likely that our electricity will need to come in the main from a mix of renewable sources, nuclear power and coal and gas with carbon capture and storage. If demand for electricity does increase, the corresponding required increase in generating capacity will depend on the success with which these sources of low carbon electricity supply

A ground source heat pump installed in a home in Wales

(some of which are intermittent) can be matched with evolving demand.

Our current installed electricity generating capacity is about 80 GW (about 75% of which comes from fossil fuels – see chapter 3). If, for example, we electrify much of our transport and heating, our demand for electricity in 2050 could be 50% higher than it is today, making it possible that electricity could account for half of our overall energy use (see chart 1). As well as nuclear power, renewable energy such as wind, and coal and gas generation with carbon capture and storage, the final generating mix could also include contributions from other low carbon sources. These include wave, tidal, solar and geothermal power, distributed and micro-generation and combined heat and power stations (potentially burning sustainable biomass), as well as technologies that are not yet developed.

If it is not possible or desirable to match low carbon electricity supply to such prospective increases in demand, then we would need to either reduce our energy demand further, or make greater use of alternative sources of energy. Bioenergy is a possible alternative in many sectors, if it is possible to address concerns about sustainability, particularly regarding biofuels – issues which are explored more fully in the next sections.

Figure 3
Our energy system in 2050 could look substantially different

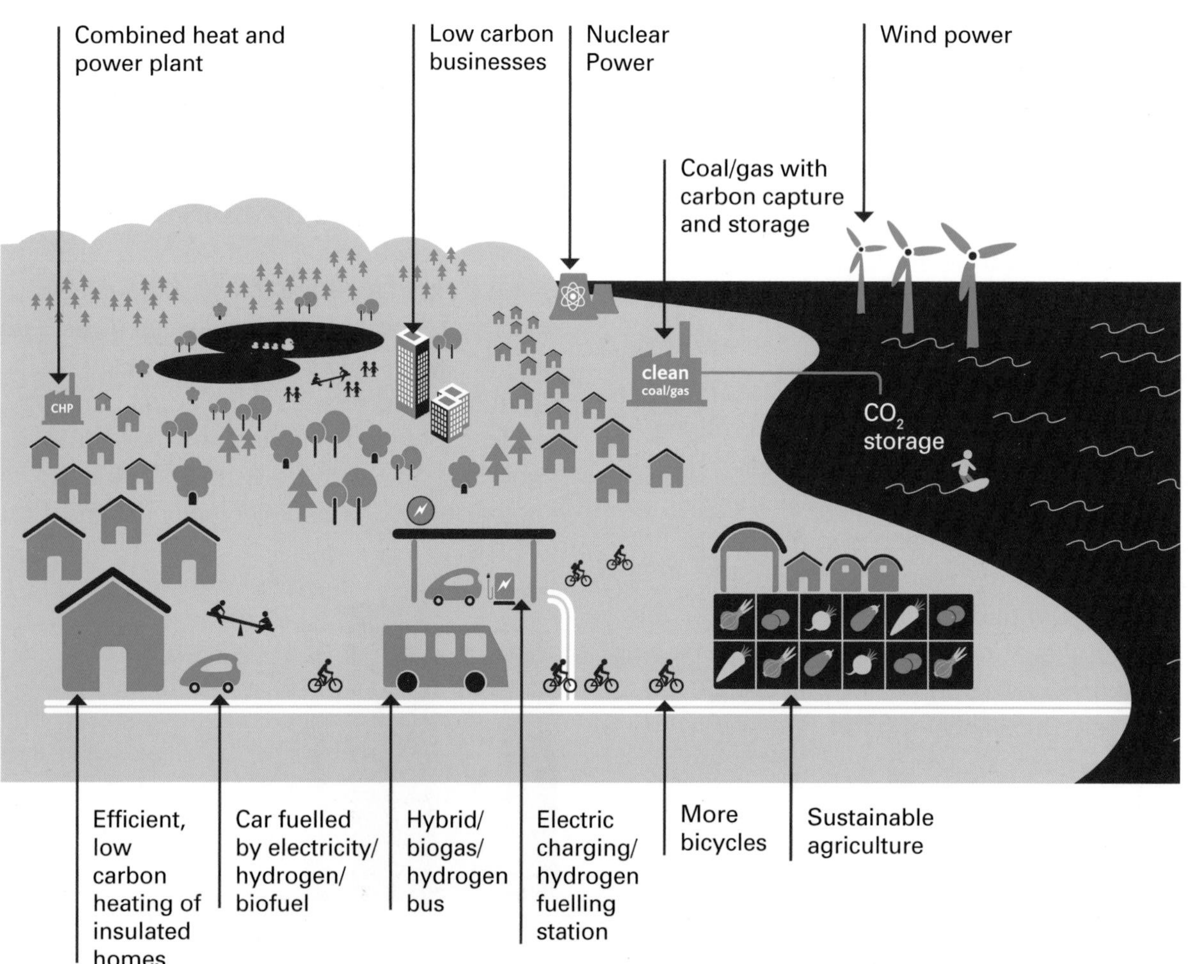

Managing areas of uncertainty

There is inevitably uncertainty about how the energy mix will evolve to 2050. Therefore, in designing a future energy roadmap, the Government will manage this uncertainty by ensuring that we build a framework robust to change, combining flexibility with appropriate regulatory and market stability.

To illustrate this, four key areas of uncertainty that need to be addressed on the path to 2050 are set out below.

The balance between reducing demand and expanding low carbon supply of energy

In the longer term, we will need to establish the best balance between efforts to reduce energy demand and the challenges and costs associated with increasing our supply of low carbon energy.

Reducing overall energy demand can potentially be very cost-effective. However, there are real and practical constraints to what may be achievable on the ground. These include:

- The scale of change we are all prepared to see in the way our homes look and are built.
- The physical constraints and engineering challenges in moving to new technologies, such as installing large numbers of ground- and air-source heat pumps, or setting up district heating systems.
- The commercial and lifestyle changes businesses and people are willing to make.

On the other hand, increasing the supply of low carbon energy has its own challenges, as discussed in the next sections. Therefore, through further work we will need to look closely at the feasibility, cost and acceptability of further reducing demand and increasing low carbon supply and understand the choices involved in balancing these.

Establishing the role of bioenergy in providing low carbon energy

Many scenarios show an increase in the use of electricity to help with cost-effective decarbonisation. However, in some cases this is offset by increasing use of bioenergy in the heat and transport sectors (see box 2).

Using sources of bioenergy could help solve some of the difficulties of using more electricity as it could provide an alternative renewable source of energy to decarbonise the heat and transport sectors. As an example, injecting sustainable biogas into the gas grid could provide an alternative to some electrical heating systems (or could supplement them on very cold days to reduce the peak in electricity demand). In *The Potential for Renewable Gas in the UK* (2009) a preliminary estimate from National Grid suggests that this could supply almost 20% of the UK's household gas needs by 2020.

In practice, we are likely to use bioenergy alongside low carbon electricity as part of our energy mix. However, there are constraints that need to be considered. For example, there have been concerns about the sustainability of some current bioenergy technologies, such as first generation biofuels, and debate over the greenhouse gas savings they can achieve. Some doubt that globally we could source the required quantities of biomass and biofuels sustainably to provide energy on a very large scale, particularly if international demand increases beyond current projections.

The biomass heating plant at Bluestone holiday park Pembrokeshire, Wales

However other analysis suggests there will be enough sustainable biomass available globally to meet our projections for bioenergy until at least 2020.[2]

If the UK met 20% of its future energy needs (for heat, electricity and transport) from bioenergy we would need to import much of this from sources abroad; we would not be able to grow this amount in the UK. The cost and feasibility of this would depend on international supply of and demand for these materials, including competition for alternative sources of renewable materials for construction, plastics and industrial chemicals and the need to maintain global food supplies.

Relying on bioenergy to deliver more than set out in the Government's *UK Renewable Energy Strategy* (2009) would require us to maximise the use of technologies such as combined heat and power and to develop advanced technologies for biofuels (including novel sources such as algae). Technologies such as anaerobic digestion and gasification would need to become commonplace. It might also be necessary to focus bioenergy use in areas where decarbonisation by other means will be difficult, such as aviation. The Government needs to continue to support the development of these options to ensure that, if viable, they are brought forward (see the Government's *Low Carbon Transport: A Greener Future* (2009) and *The UK Renewable Energy Strategy* (2009)). Crucially, for bioenergy use to reach its full potential we must ensure that clear and effective international standards on sustainability are put in place.

Technology paths

The mix of low carbon technologies and energy sources used in 2050 is also unclear, particularly in heat and transport. In time it is possible that different forms of transport will use different energy sources: for example, aviation and heavy goods vehicles may run on biofuels or still rely on fossil fuels, while lighter vehicles may use mostly electricity or hydrogen. The eventual mix has considerable implications for the UK's infrastructure, as well as other sectors such as electricity generating capacity.

Similarly, our homes and buildings could plausibly be heated, or cooled, using a variety of low carbon technologies: heat pumps, boilers burning sustainable biomass or biogas, or perhaps using the captured waste heat from electricity generation. How this energy mix evolves will have significant impacts on the other elements of the energy system.

It is also impossible to predict the precise mix of electricity generating technologies we will use in 2050. The roadmap to 2050 needs to be sufficiently flexible to adapt to technological developments in any sector.

2. For example, see E4tech's Biomass supply curves for the UK (2009)

Box 2
What might be the role of energy from recent biological sources in a low carbon energy system?

Bioenergy comes from a wide range of renewable sources including wood, energy crops and organic wastes such as sewage and waste. It is also used in a number of forms.

There are many ways of using bio-energy:

Electricity generation: We can use dedicated biomass power stations, or can co-fire coal with biomass in power stations (and potentially with carbon capture and storage technology as well). This provides flexible generating capacity to balance intermittent sources like wind, which helps to reduce overall emissions. When used in combined heat and power plants, the waste heat is captured and used to heat local businesses and homes.

Heating: The simplest use of biomass is through direct burning in biomass boilers, which can be used to heat our homes or our water. We can also use biogas or synthesis gas to produce heat.

Transport: Our cars can already run on a blend of biofuel and conventional fossil fuel. Currently, two primary biofuels are in commercial production: bioethanol and biodiesel. In the near future, biomass-to-liquid technologies are likely to be used to produce biofuels.

Only sustainably produced biomass should be used. The Government needs to work to ensure that effective mechanisms are in place that address potential negative impacts on biodiversity, land use, food prices or other sustainability concerns - domestically or internationally.

Sources of residual emissions

Finally, there remains uncertainty about the scale at which we will still need to use energy sources that create emissions. Any use will have to be consistent with the UK's overall targets, emissions reductions in other sectors and the extent to which we can trade emissions in 2050. Examples of energy-related emissions that might be hard to eliminate can be found in transport (perhaps from aviation or heavy vehicles, or in cars for longer journeys), heating (perhaps from burning gas on very cold days) and power (from residual emissions from carbon capture and storage, or perhaps from combined heat and power units, or gas-fired power plants supplying peak demand). This is another area of uncertainty that our long-term planning will need to accommodate.

Figure 4
Changes in one sector could affect changes in other sectors

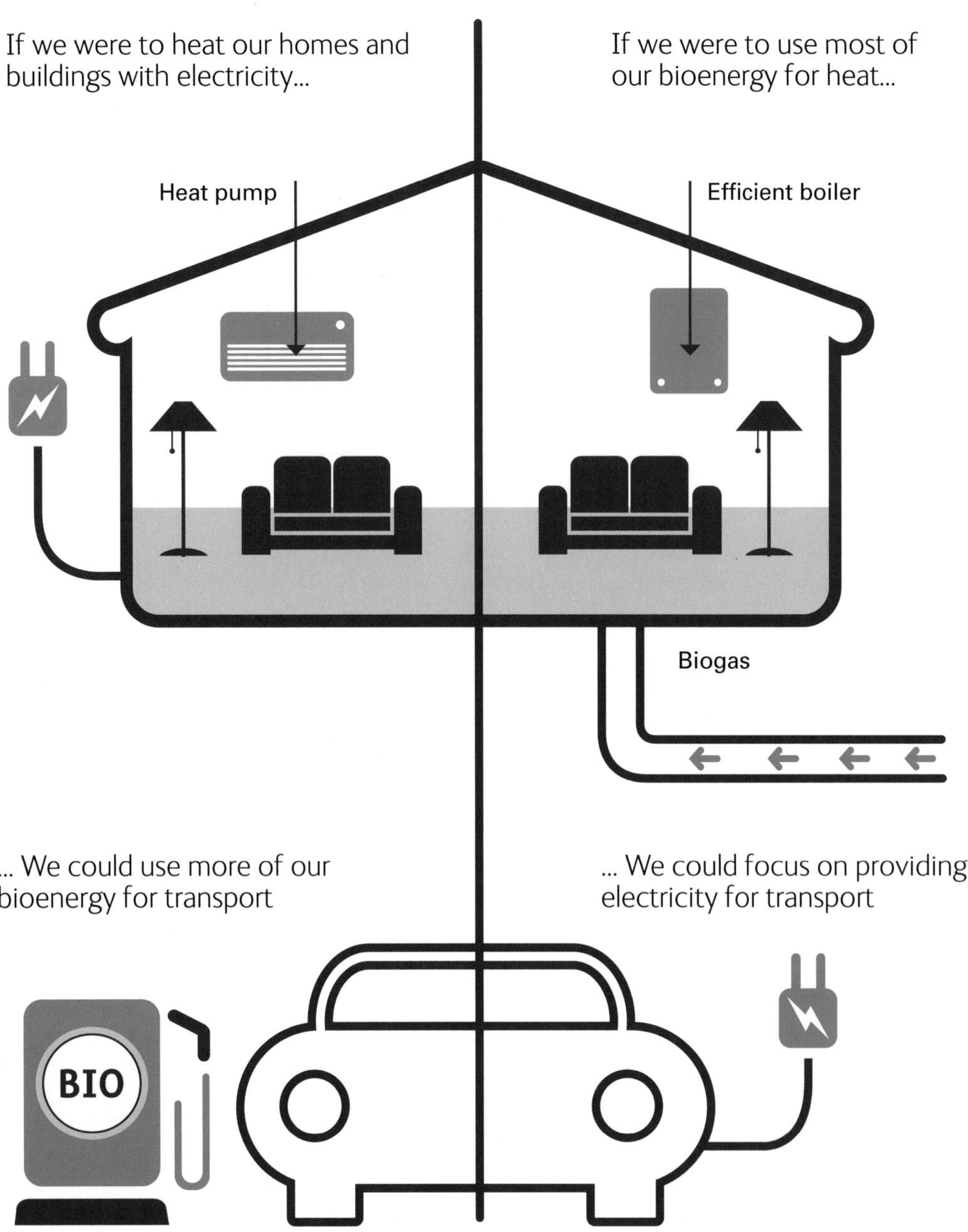

Source: Department of Energy and Climate Change

Practical challenges

Changing our energy system will take place over decades, and will be driven by an appropriate market and regulatory framework to encourage private sector investment, based on a carbon price. However, there are a number of practical considerations that the Government and industry need to keep under review to ensure that there is sufficient strategic direction to keep us on a path that will meet our long-term emission targets:

- Planning and enabling timely investment:
  - Delivering sufficient financial investment, and ensuring the attractiveness of the UK as a place to invest.
  - Ensuring planning policies support the development and installation of low carbon technologies.
  - Taking advantage of the replacement/refurbishment schedules of existing plants and infrastructure.
- Delivering the engineering challenges of building a low carbon energy system of this scale:
  - Meeting the physical and supply chain challenges of building new, reliable electrical generating capacity and other energy infrastructure at this scale, particularly in the face of likely international competition for these capabilities.
  - Developing or upgrading infrastructure as it becomes necessary. As well as the electricity grid (transmission and distribution networks), this could include networks for transporting and storing captured carbon, systems for managing nuclear waste, hydrogen or electric vehicle fuelling/charging networks and community heat systems.
  - Matching evolving sources of demand for energy with new sources of supply in an efficient and practical manner.
  - Building the necessary skills base.
  - Developing technologies that we are relying on or that we may need to plan for.
- Dealing with broader societal impacts:
  - Consumer and business acceptance of new ways of travelling, heating buildings and using appliances.
  - Impacts on the local environment of different technology paths.
  - The impact on the least well off of moving to a low carbon society.

The transition to low carbon transport could require substantial additional infrastructure, such as charging points for electric cars

Securing change in the long term

This Transition Plan sets out a clear plan that put us on the right path to 2020 and beyond. Given the scale of the longer term changes required and the time it takes to achieve them, the Government, industry and stakeholders need to examine how to continue to drive cost-effective progress towards 2050, whilst maintaining secure energy supplies, maximising economic opportunities and protecting the most vulnerable.

Therefore the Government will develop a strategic roadmap to 2050 by spring 2010, working closely with industry and wider stakeholders. The roadmap will help to build consensus between the Government, industry and the public on the scale and nature of the changes we need to see and the issues that need to be addressed, enabling major infrastructure and capacity - building investments to be made.

In autumn, the Committee on Climate Change will provide further analysis of the pathway through 2030 to 2050. The Government will work with the Committee, taking its analysis and recommendations into account when developing the roadmap to 2050.

Chapter 9

Further action in Northern Ireland, Scotland and Wales

Summary

Action on energy and climate change is taking place across the whole of the UK. In Northern Ireland, Scotland and Wales the devolved administrations are responsible for some areas of energy and climate change policy, with differences in responsibility between each administration. This part of the plan sets out the ambitions and actions that the devolved administrations in Northern Ireland, Scotland and Wales are taking and how this is contributing to developments across the whole of the UK.

Each devolved administration has set out key targets and ambitions that reflect its responsibilities and circumstances:

- Ireland: Through its *Programme for Government 2008-2011*, the Northern Ireland Executive has committed to reduce emissions by 25% on 1990 levels by 2025. The Executive and Northern Ireland Assembly have also consented to the extension of the Climate Change Act 2008 to Northern Ireland and are contributing to UK carbon budgets and targets.
- Scotland: The Climate Change (Scotland) Bill as passed sets a mandatory target to achieve an 80% reduction in 1990 levels of Scottish greenhouse gas emissions by 2050 and creates a statutory framework committing the Scottish Government to securing this reduction. The Scottish Government is committed to delivering the highest possible interim 2020 target based on expert advice from the Committee on Climate Change, and the Bill as passed sets the 2020 target to 42%.
- Wales: The Welsh Assembly Government in *One Wales - A progressive agenda for the government of Wales*, made a commitment for Wales to reduce annual greenhouse gas emissions by 3% each year in areas of devolved competence by 2011 and to set out specific sectoral targets in relation to residential, public and transport areas. Through the Climate Change commission for Wales, the Assembly Government is exploring the implications of more ambitious emission reduction scenarios of 3%, 6% and 9%. The Assembly Government also wishes to see Wales by 2025 producing more electricity each year from renewables, especially from marine resources, than the electricity consumed annually by the nation.

Action to tackle climate change is happening across the UK

Some matters which relate to climate change and energy policy in Northern Ireland, Scotland and Wales are the responsibility of the devolved administrations, and therefore decisions on these matters are made in the light of each administration's particular circumstances.

Which areas are devolved and which are not vary in each case, but in general terms each devolved administration has programmes on low carbon economic development, fuel poverty, energy efficiency, and environmental, agricultural and rural policy. Energy is particularly complicated; for example, Northern Ireland's energy system is closely linked to the Republic of Ireland's. This chapter sets out the approach being taken by each of the devolved administrations in relevant policy areas outlining just some of the range of activity being undertaken in each sector of the economy.

Northern Ireland

Through its *Programme for Government 2008-2011*, the Northern Ireland Executive has committed to reduce emissions by 25% on 1990 levels by 2025. The Executive and Northern Ireland Assembly have also consented to the extension of the Climate Change Act 2008 to Northern Ireland and are contributing to UK carbon budgets and targets. Northern Ireland is also participating in a number of relevant EU and UK energy and climate change policies. Through the Climate Change Act 2008, Northern Ireland is committed to the development of a UK risk assessment on the impacts of climate change and this will inform the development of a Northern Ireland adaptation programme.

Scotland

Through its *Government Economic Strategy* the Scottish Government has committed to reducing Scottish emissions. The Climate Change (Scotland) Bill as passed sets a mandatory target to achieve an 80% reduction on 1990 levels of Scottish greenhouse gas emissions by 2050 and creates a statutory framework committing the Scottish Government to securing this reduction. The Scottish Government is committed to delivering the highest possible interim 2020 target based on expert advice from the Committee on Climate Change. The interim target provision in the Bill as passed sets the 2020 target to 42%, requiring Scottish Ministers to seek expert advice at the earliest opportunity on what the highest achievable 2020 target should be. The Bill provides order making powers to revise the 2020 target to match the expert advice – this means the target could go up or down.

The Bill as passed commits the Scottish Government to publishing and reporting on an *Energy Efficiency Action Plan* covering all sectors. The draft Action Plan will be published for consultation later in the summer of 2009. The Bill as passed also includes a number of provisions to support waste reduction and recycling, in line with Zero Waste principles and improvements to the energy performance of the existing building stock. The *Climate Change Delivery Plan: Meeting Scotland's Statutory Climate Change Targets*, setting out strategic options for delivering future emissions cuts in Scotland was published in June 2009. The Scottish Government is also committed to assessing the impact on carbon emissions of its own spending and is developing a *Climate Change Adaptation Framework* to make sure Scotland can adapt to the impacts of climate change which are already being felt.

Wales

The Welsh Assembly Government in *One Wales - A progressive agenda for the government of Wales*, made a commitment for Wales to reduce annual greenhouse gas emissions by 3% in areas of devolved competence by 2011 and to set out specific sectoral targets in relation to residential, public and transport areas. The focus is on action in Wales, which will be led by the Assembly Government, but with a strong emphasis on all sectors playing a role in reducing emissions and adapting to the impacts of climate change.

Achieving the 3% annual reduction will be Wales' contribution to the UK carbon budgets and targets. The Assembly Government's policies and programmes will be set out in a Climate Change Strategy for Wales, which is being developed in two stages. The first stage – *Climate Change Strategy – High Level Policy Statement consultation*, published at the beginning of the year, sets out the definitions of the targets and the broad areas where Wales will focus action to tackle both the causes and consequences of climate change. The second stage, *Climate Change Strategy for Wales – Programme of Action* was published in June 2009. This sets out how the Assembly Government is going to meet its One Wales target, detailing the action to be taken to reduce emissions and adapt to the impacts of climate change. The final Climate Change Strategy for Wales, bringing together the results of the two consultations, will be published at the end of the year and be informed by the Commission's scenario work. The Assembly Government will also publish a comprehensive Energy Strategy, bringing together the Wales Energy Route Map, National Energy Efficiency and Saving Plan, Marine Energy Statement and Bio-Energy Action Plan.

Wales is committed to reduce carbon dioxide emissions by 3% per year by 2011

Transforming our power sector

Northern Ireland

In Northern Ireland, the proportion of electricity consumption generated from renewable energy sources has doubled since the introduction of the Northern Ireland Renewables Obligation in 2005. Currently, indigenous renewables generation represents 7% of total electricity consumption and is on course to meet the 12% target set for 2012 – mainly from onshore wind power. The Department of Enterprise, Trade and Investment (DETI), which has responsibility for energy matters in Northern Ireland, has

Renewable electricity has doubled in Northern Ireland since 2005

recently published a draft Strategic Energy Framework for public consultation. This has a focus on renewables and proposes a new target for 2020 that will contribute to the overall UK renewable energy target of 15%.

DETI is working with other departments and agencies to develop non-wind renewables in order to diversify the renewable energy mix. As part of this, the Department is developing a Strategic Action Plan for offshore wind and marine renewables (tidal stream and wave). This plan is currently the subject of a Strategic Environmental Assessment (SEA) and will be issued for consultation in Autumn 2009 alongside the SEA's environmental report. It is also proposed to consult in Summer 2009 on the first Cross Departmental Bioenergy Action Plan.

Increasing levels of renewable energy generation will also require significant investment in the electricity grid in Northern Ireland. A programme of grid strengthening is being developed.

Scotland

Scottish Ministers are committed to promoting energy from a wide range of renewable sources, and have a target that 50% of electricity generated in Scotland, as a proportion of demand, should come from renewable sources by 2020, with an interim milestone of 31% by 2011. Much of the developer activity to date, driven by the Renewables Obligation Scotland, has been focused on onshore wind. In common with the rest of the UK, the introduction of banding to give different levels of support through the Renewables Obligation Scotland for different renewable energy technologies is expected to bring on a wider range of technologies. The Scottish Government continues to provide additional support to small- and micro-scale renewables through its Communities And Renewable Energy Scheme and Energy Saving Scotland – Home Renewables service.

Scotland is making progress towards 50% of its electricity coming from renewable sources

Wales

The *Renewable Energy Route Map*, published for consultation in February 2008, sets out Wales' agenda for exploiting its exceptional renewable energy resources. Currently some 360MW of onshore wind energy is operational in Wales, with 60MW of off shore wind development at North Hoyle and a further 100MW of offshore wind development currently under construction. The second largest offshore windfarm in the world, 750MW at Gwynt y Mor, received consent in December 2008 and, in addition to the enormous industry efforts to deliver Wales, ambitious on-shore wind targets, more than 5000MW of further offshore wind is in prospect. In addition 150MW of hydro power is operational in Wales and in north Wales two major pumped storage stations continue to play an important role. There is also substantial co-firing of biomass in Wales, coal fired power stations and a number of significant biomass projects, including energy from waste, in prospect.

The Welsh Assembly Government is working with National Grid Transco on necessary electricity grid enhancements and is taking action to remove barriers to investment. *Planning Policy Wales*, published in March 2002, sets out the Welsh Assembly Government's planning policies and climate change is a pivotal factor. In 2006 the

Assembly Government consulted on a range of proposed changes to further increase the emphasis on tackling climate change in national planning policy. These policies have been further updated since the One Wales 3% target was adopted in 2007. To implement policies and technical advice, the Assembly Government has also funded a series of training seminars on planning for climate change. Against the background of the joint Assembly Government / UK Government / South West England Severn tidal power feasibility study, Wales' renewable energy route map and the associated Welsh Assembly Government bio-energy action plan and Ministerial marine energy statement of intent, investments in renewable electricity production of up to £40 billion are in prospect.

Similarly, as well as encouraging community renewable electricity developments, the Welsh Assembly Government will be seeking to ensure that significant amounts of renewable heat are produced in Wales both from the use of biomass from sustainable sources and from community scale renewables and micro-generation systems.

Transforming our homes and communities

Northern Ireland

In Northern Ireland, all new build social housing is required to meet Level 3 of the Code for Sustainable Homes, and housing associations are offered incentives to exceed these standards. There is currently a target for all social housing in Northern Ireland to meet the Decent Homes standard by 2013. Existing housing across tenures has become substantially more efficient and is improving, thanks to schemes such as Warm Homes, Cosy Homes and the Heating Replacement Scheme.

Action is taking place across the UK to cut emissions from our homes

Under the Energy End Use Efficiency and Energy Services Directive, Northern Ireland has implemented, as part of the UK Energy Efficiency Action Plan, a 1% year on year energy savings target. A *Northern Ireland Energy Efficiency Action Plan* is currently being developed and should complete towards the end of 2009. This Action Plan will identify energy efficiency measures and opportunities that can be implemented during the next two years. The Department of Enterprise, Trade and Investment (DETI) has now signed voluntary agreements with all of the major energy suppliers in Northern Ireland (electricity, gas, oil, coal and biomass) to provide energy efficiency advice and information, collection of energy data, monitoring of targets, and provision of energy audits.

The DETI has now completed a consultation on better energy billing and metering. As a result, new regulations[1] have been introduced to ensure electricity and gas suppliers provide a year's historical consumption data to all domestic customers. The DETI continues to work to improve public engagement on

1. The Electricity and Gas Billing Regulations (Northern Ireland) 2009.

sustainable energy issues, and will produce a sustainable energy marketing plan by the end of 2009.

Scotland

The Scottish Government has supported energy efficiency improvements to be made in homes and wider communities.

The Energy Saving Scotland advice network provides a one-stop-shop offering support on a range of sustainable living issues, including energy efficiency, microgeneration, transport and waste. The network is managed by the Energy Saving Trust. The Scottish Government is introducing an area-based *Home Insulation Scheme*, to increase the take up of energy advice and insulation measures in selected areas, to reduce emissions, tackle fuel poverty, reduce household bills and sustain jobs. This is supported by £15 million of investment in 2009-10, with matching investment being sought from other sources.

Scotland has also implemented the European Directive on the Energy Performance of Buildings (EPBD), with energy performance certificates required for both domestic and non-domestic buildings that are newly built, offered for sale or rental, and for public buildings. Scotland also requires inspections of larger air-conditioning systems under the EPBD.

Scottish building regulations are devolved. In 2007, an expert panel advised on staged improvements to standards in 2010 and 2013, with the goal of net zero carbon buildings by 2016-17, if practical. Energy standards are the most demanding in the UK and a consultation on measures to reduce emissions by a further 30%, with effect from October 2010, is in progress.

Scotland introduced a more holistic approach to tackling fuel poverty with the new Energy Assistance Package introduced in April 2009. This builds on the success of the Scottish Government's previous Central Heating and Warm Deal programmes, but provides more help for a wider range of households. Clients are offered benefit and tax credit checks, tariff checks, energy efficiency advice, access to cavity and loft insulation measures through CERT providers and, for those households most vulnerable to fuel poverty, access to grants for enhanced insulation and heating improvements.

Wales

Supporting people and communities to reduce their carbon footprint of their homes is one of the key themes of the Welsh Assembly Government's emerging Climate Change Strategy. Consequently, the Welsh Assembly Government has set out its aspiration that all new buildings in Wales be zero carbon from 2011. It also requires all new housing that it influences through grant funding, investment and land disposals to meet at least Level 3 of the Code for Sustainable Homes, moving as quickly as possible to Level 4. For social housing, a minimum energy efficiency rating must be achieved by all existing social housing within Wales, by 2012.

The National Housing Strategy sets out the Welsh Assembly Government's long-term vision for housing. This and the *National Energy Efficiency and Savings Plan* aims to reduce Wales' greenhouse gas emissions, reduce fuel poverty and support economic development. It proposes actions to help all householders, and seeks to help communities work together.

The Welsh Assembly Government's *Home Energy Efficiency Scheme* has helped to reduce household emissions. More than 100,000 householders have been helped to save money since 2000. The Welsh Assembly Government has consulted on targeting funding at the most inefficient properties and those people most in need of support.

It is also running a major climate change campaign, encouraging people to measure and reduce their emissions.

Climate change is fully integrated into the Welsh Assembly Government's national planning policies, and *Planning Policy Wales* has been updated to increase the emphasis on tackling climate change.

Looking ahead, the Assembly Government is proposing to develop an area-based approach to domestic energy efficiency improvements. The approach will include projects to tackle fuel poverty using new technology, stimulating the supply chain and creating job opportunities including through the Heads of the Valley Low Carbon Zone. It will support community scale energy generation projects, and make funding available for farm renewable energy projects. The Welsh Assembly Government will work with the Energy Saving Trust and Carbon Trust to provide more advice and support, and will promote the use of waste heat and improved energy efficiency in businesses.

Transforming our workplaces and jobs

Northern Ireland

Incentivising and supporting businesses to reduce emissions

Invest NI, which provides Government support to encourage investment in Northern Ireland and help new businesses grow and compete internationally, runs resource efficiency activities. These are aimed primarily at increasing the productivity and competitiveness of businesses through more efficient use of materials, water and energy. They comprise funding of the local delivery of nationwide programmes, the Carbon Trust, loan schemes to help fund the deployment of sustainable energy equipment and practices, Envirowise and the *National Industrial Symbiosis Programme.*

Wave striking Limpet 500, the world's first commercial-scale wave power station. Limpet generates 500 kilowatts of electricity, enough to power 300 homes (Islay, Scotland)

Leading the way in the public sector

Northern Ireland's Public Sector Energy Campaign sets carbon reduction, energy efficiency and electricity consumption targets for public sector estate buildings in Northern Ireland, which are measured annually in published reports.

To support delivery of these targets, the Department of Finance and Personnel currently makes available £2 million annually in grant assistance via the Central Energy Efficiency Fund. Successful projects may receive a grant that will cover up to 100% of the capital cost of the project.

Supporting UK businesses to make the most of new economic opportunities

Invest NI will work with any manufacturing or tradable services business in Northern Ireland which has the potential and ambition to export, to improve its productivity and to become more internationally competitive. An Invest NI strategy to support businesses to take advantage of the opportunities offered by a low carbon economy is currently being developed.

The strategy aims to position Northern Ireland as a leading region in renewable energy development, securing greater business for clients in renewable energy supply chains. Initiatives aligned to the strategy include awareness raising and publicising of business opportunities; encouraging and facilitating business clustering and networking; promotion of the region as an FDI candidate for companies in renewables; identification and promotion of regional strengths in low carbon technologies; participation in the UK Government's Renewables Deployment Forum; support of renewables trade missions overseas; and encouraging businesses and the local universities to establish centres of competence in renewables.

Scotland

Incentivising and supporting businesses to reduce emissions

Business and industry in Scotland is playing a key role in tackling climate change, with many enterprises taking steps to improve their energy efficiency and cut waste, as well as seizing the business opportunities presented by the low carbon economy.

The Scottish Government is supporting the work of the Climate Change Business Delivery Group, which shares ideas, challenges Scottish business to do more to tackle climate change, acts as a source of inspiration and information for others in the Scottish business community, and influences policy and practice.

To promote awareness, the Scottish Government is funding *Scottish Business in the Community (2008-2010)* to support the May Day Network which encourages businesses to make climate change pledges and take energy efficiency measures.

A further £2 million has been invested in the loan scheme for small and medium sized enterprises to improve their energy efficiency, specifically for microgeneration support. This takes the total amount invested in this scheme to £5 million.

Microgeneration equipment has been exempted from rating valuation for the purposes of non-domestic rates. The measure removes a potential disincentive for businesses to invest in microgeneration equipment – with a capacity of up to 50kW or 45kW thermal.

The Scottish Government funds the Carbon Trust to provide technical energy efficiency advice and change management support to large and energy-intensive businesses. This may include on-site energy audits or a more bespoke service to suit the needs of the business. This support has resulted in significant carbon and cost savings for industry. The Scottish Government also funds the Envirowise programme, to provide advice to business on resource efficiency, waste prevention and the sustainable use of water.

Leading the way in the public sector

In Scotland, all local authorities have signed Scotland's *Climate Change Declaration* and committed themselves to take action, in partnership with the Scottish Government. The Carbon Trust's *Public Sector Carbon Management Programme* has proved very successful in Scotland. The impact of the Carbon Management Programme continues to grow year-on-year with the latest group of plans estimated to save over 150,000 tonnes of CO_2 over the next five years.

The Scottish Government continues to support the public sector through the Central Energy Efficiency Fund, which has seen £20 million provided to Scottish Local Authorities, the NHS Scotland and Scottish Water. Through interest free loans, these bodies use the scheme for capital investment in energy efficiency projects and, as of 2009, also for renewables technologies. In 2008 the Scottish Government also awarded a further £4 million, managed through Salix Finance, for the further and higher education sector in Scotland.

Public sector spending on goods and services across Scotland amounts to approximately £8 billion per year. The Scottish Government issued guidance to the public sector in October 2008 recommending the use of the Buy Sustainable - Quick Wins which are a range of detailed specifications for commonly purchased goods that can be adopted to deliver sustainable outcomes.

Supporting businesses to make the most of new economic opportunities

Major investment plans in the Scottish renewable sector, remain on track. Onshore wind in Scotland also remains resilient, illustrated by the announcement from ScottishPower Renewables that construction of Europe's largest onshore wind farm at Whitelee, near Glasgow, has now been completed ahead of schedule.

New technologies such as offshore wind are creating new opportunities for UK businesses

Wave and tidal energy still remains an area of huge potential growth. With full applications for the first round of the Crown Estate's leasing programme focused on the Pentland Firth due to be submitted in mid May. This development area includes the site of the first test centre for wave and tidal technology in the world, the European Marine Energy Centre (EMEC) in Orkney. The potential to unlock the resources in the Pentland Firth, which is estimated to contain six of the top ten sites for tidal developments in the UK, will promote both leadership in this industry along with valuable new jobs in Scotland.

The Crown Estate has identified 10 potential sites for offshore development within Scottish Territorial Waters with the potential to generate 6.4GW of renewable electricity.

Wales

Incentivising and supporting businesses to reduce emissions

Supporting businesses and the public sector in Wales to reduce their carbon footprint is a key theme of the Assembly Government's Climate Change Strategy. As part of its Green Jobs Strategy the Assembly Government is investing in a new component of its *Flexible Support for Businesses* to respond to the challenge of climate change. *Flexible Support for Business Environment and Sustainability* will enable businesses to access the best possible knowledge on sustainable and cost effective business practices and contribute towards Wales achieving a low carbon economy. In addition, the Welsh Assembly Government will be providing enhanced support and advice to small and medium sized enterprises through the Carbon Trust.

The Assembly Government recognises that the public sector has a critical role to play in visibly demonstrating leadership on climate change action and this is a key theme in the emerging Climate Change Strategy. The focus of action will be to reduce the carbon footprint of public sector organisations across Wales, supported by a comprehensive programme to educate and raise awareness of climate change across the sector and by providing the skills needed by a green workforce.

Supporting UK businesses to make the most of new economic opportunities

The Welsh Assembly Government published its final Green Jobs Strategy in July 2009. It forms part of a much broader range of measures designed to create a sustainable economy built on the firm foundations of sustainable businesses, sustainable technologies and sustainable employment. The strategy is organised around three priorities: supporting business, fostering innovation and technology, and investing for a more sustainable economy.

The Green Jobs Strategy will play a key part in shaping and driving the business opportunities associated with a move to a low carbon, low waste economy, including seizing on the local supply chain opportunities which will be created by the billions of low carbon electricity investments in Wales which these climate change and energy policies will be encouraging.

These developments will make an important contribution to the way in which Wales deals with the current economic situation and places itself in a robust position to take advantage of future opportunities.

Transforming transport

Northern Ireland

Road transport is the largest source of greenhouse gas emissions in the North of Ireland, accounting for 29% of total carbon emissions. Delivering reductions in greenhouse gas emissions from road transport, while ensuring the provision of transport arrangements that meet economic and social needs, presents a significant challenge. Reconciling these potentially competing priorities is a key objective.

The Northern Ireland Executive and Assembly recognise the real need to reduce greenhouse gas emissions across the transport sector to meet the challenge of climate change. The Programme for Government, in line with the Sustainable Development Strategy, has set an ambitious target for a reduction in local emissions of greenhouse gases. To contribute fully to realisation of the target, a 51% reduction is required on 2006 greenhouse gas emissions from road transport by 2025.

The immediate priority is to establish the baseline greenhouse gas emissions from transport and the options for reduction. The *Regional Transportation Strategy* is currently being reviewed to enhance the suite of policy and operational measures needed to ensure that future transport arrangements are more sustainable.

The Executive and Assembly's current investment plans – set out in the *Investment Strategy for 2008-2018* – envisage a significant level of investment in public transport, including the Belfast Rapid Transit project. Planned investment in buses, trains and related facilities aims to promote increased utilisation of public transport and reduce dependency on the private car, thereby contributing to climate change targets. To facilitate progress in this area, action will continue to promote behavioural change and more sustainable modes of travel.

Scotland

The Scottish Government's *Climate Change Delivery Plan: Meeting Scotland's Statutory Climate Change Targets*, shows that strong demand management measures will be needed up to 2020. Significant uptake of low carbon vehicles, coupled with smarter travel, should contribute towards a 50% reduction in land transport emissions by 2030. Further vehicle changes, coupled with the potential development of alternative fuels should make a 90% reduction in land transport emissions feasible by 2050.

Scotland's *National Transport Strategy* commits to develop a carbon account for transport, to monitor progress, and show which transport policies are forecast to have the most impact.

The Scottish Government will consult in Summer 2009 on how to accelerate the development and uptake of low carbon vehicles, and possible targets to be set. Work is underway to benchmark the Scottish public sector fleet, and to identify potential for greener vehicle procurement.

Smarter Choices, Smarter Places is a partnership project which makes up to £15 million available, over three years, for seven Local Authorities to improve public transport services, walking and cycling infrastructure, and roll out intensive marketing and awareness campaigns.

The Scottish Government's draft Cycling Action Plan for Scotland is being consulted on over Summer 2009 and proposes a target of 10% of all journeys made by bike by 2020. The Scottish Government also makes funding available to organisations, including Local Authorities, to promote active travel initiatives.

The Scottish Government funds the Energy Saving Trust to engage individuals and organisations. The Trust and the Energy Saving Scotland Advice Network promote eco-driving. The Trust provides free, bespoke advice on fleets (Green Fleet Reviews) and Travel Plans. Additional support on Travel Plans and Smarter Choices initiatives are provided by Travel Plan officers in each Regional Transport Partnership.

The Scottish Government is reviewing the bus subsidy, paid as Bus Service Operators Grant, to link it more closely to reduced environmental impact.

The Scottish Government is investing over £500,000 in 2008-11 in extending the Freight Best Practice Programme into Scotland as part of its support for the freight industry. As part of the UK-wide Sustainable Rail Programme, the Scottish Government is planning a rolling programme of electrification of Scotland's railways and greater efficiency from the whole of the rail sector.

A major review of Scottish ferry provision is currently underway, to identify options for significant emission reduction measures. Alternative fuels and innovative vessel design measures are being considered as part of the review.

Emissions from domestic and international aviation are included in the Scottish Government's Climate Change Bill.

Wales

Supporting emission reductions across the transport sector is a key theme of the Assembly Government's emerging Climate Change Strategy. The Wales Transport Strategy was published in 2008 and the National Transport Plan was published in July 2009. Regional Transport Plans are being developed and it is anticipated that the final regional transport plans will be in place by December 2009.

The Assembly Government and partners are working to encourage people to make choices that will help reduce their carbon footprint and a reduction in car usage. This includes enhanced provision for walking and cycling, as well as public transport, park and ride, high occupancy vehicle lanes and the promotion of eco-driving techniques. The Welsh Assembly Government has Action Plans and targets to support this and has announced plans for developing Sustainable Travel Towns in Wales. These will target a series of focused 'smarter choice' interventions. The Assembly Government funds the Energy Saving Trust to provide consumer travel advice and advice to fleet operators.

The Welsh Assembly Government is helping people to choose to reduce transport emissions

The Welsh Assembly Government is also planning to establish a centre for inter-modal freight logistics, which would create maximum efficiencies with the freight sector.

Transforming farming and managing our land and waste sustainably

Northern Ireland

Work is underway to explore emission reduction options for agriculture in Northern Ireland. This will help develop a vision for local agriculture's contribution to climate change targets by 2050. The Northern Ireland Executive is already taking action in a number of areas:

The Nitrates Directive Action Plan, control of stocking levels and fertiliser use under agri-environment schemes, support for increased efficiency in manure spreading technologies, and demonstration and training programmes will all contribute to reduced agricultural emissions up to 2020.

Increasing our forests helps our efforts to cut greenhouse gases

The Department of Agriculture and Rural Development Forest Service's Strategy for Sustainability and Growth aims to double tree cover from 6% to 12% of land area by 2056. This will improve opportunities for carbon sequestration, while ongoing support of short rotation willow coppice for renewable energy will also contribute to emissions reductions.

Scotland

The Scottish Government is funding a range of research to understand the role of land use in greenhouse gas emissions and their mitigation. This includes research on the impacts, costs and benefits of biomass energy crops, the role of forestry in relation to carbon 'sinks', changes in consumer demand in the move to a low carbon economy, and understanding policy instruments to reduce greenhouse gas emissions.

In May 2008, the *Graham Report* was published by the Scottish Government. This was the output of a stakeholder group, chaired by Henry Graham, which looked at the main issues relating to agriculture and climate change in Scotland. In May 2009, *Farming for a Better Climate*, a five point action plan, was published which suggests ways to reduce greenhouse gas emissions whilst strengthening farm businesses to make them more resilient to climate change. Further work is now underway to ensure that the agricultural sector plays its part in meeting the Scottish Government targets for reducing emissions as set out in the Climate Change (Scotland) Bill as passed.

The Scottish Government is publishing a new *National Waste Management Plan for Scotland*, outlining ambitious targets on waste prevention, re-use and recycling.

The *Scottish Forestry Strategy*, published in 2006 identifies climate change as a central theme. The strategy sets out how the Scottish Government will:

- Increase awareness of how the forestry sector can help to tackle the threats of climate change.
- Ensure that Scotland's woodlands and the forestry sector meet their full potential in facilitating ecological, economic and social adaptation to climate change.
- Capture opportunities for forestry to help mitigate climate change through the use of wood resources and habitat management.
- Increase the amount of carbon locked up by Scottish forestry.

In addition, early in 2009 Forestry Commission Scotland published a climate change action plan (2009 –2011). It describes what the Commission will do to increase the contribution of Scottish forestry to the challenges of climate change and focuses on what needs to be done both as early actions and to increase future readiness. Key actions include creating new woodland to increase the potential for forests to absorb carbon dioxide, expanding the use of woodfuel as a form of renewable energy, and increasing the use of timber in place of more carbon intensive materials.

Scotland has the greatest capacity within British forests to absorb carbon dioxide from the atmosphere. This is due to the scale of forest cover in Scotland and the types of species that it includes. In 2006, forests in Scotland removed about 10.1 million tonnes of carbon dioxide from the atmosphere. Considerable potential exists to extend this capacity in future. A sustained planting programme of 10-15,000 hectares of new woodland per year could result, by 2050, in a further 5 million tonnes of carbon dioxide per year being absorbed.

New farming methods are reducing greenhouse gas emissions

Wales

The report *Sustainable Farming and Environment – Action towards 2020* recommends that action is taken by the Welsh Assembly Government to achieve carbon neutral status for agriculture by 2020.

Through Farming Connect, the Assembly Government will promote nutrient and resource management planning and best practice advice. It will ensure that adherence to the *Code for Good Agricultural Practice* occurs. The Welsh Assembly Government will use Farming Connect farm development programmes to deliver technical efficiency methods for the dairy, beef, sheep, arable and horticulture sectors to deliver emission reductions. It will also encourage farmers to take stock of farm emissions through the use of an on-farm carbon-accounting tool which is part of the new agri-environment monitoring contract.

A new Knowledge Transfer Development Programme for Climate Change – to be launched in July 2009 – will deliver advice regarding on-farm energy efficiency and renewables, soil carbon and woodland management.

The Rural Development Plan for Wales is being implemented using expenditure under the EU Rural Development Regulation for 2007-13, of which Environmental Stewardship is a key part. Addressing climate change is an integral element of this programme.

From January 2012 a new agri-environment scheme, Glastir, will be introduced and will have a strong focus on tackling climate change. This is includes promoting land management techniques that preserve soil carbon and a grant scheme to support the development of on-farm renewables.

The Welsh Assembly Government is also preparing road maps from production to consumption of red meat and dairy as a way of analysing the greenhouse gas emissions and water use at all stages of the food chain. It will then produce an action plan to reduce the emissions.

Through the Forestry Commission, the Assembly Government is responsible for the public forest estate that cover 6% of Wales' land, as well as the regulation of the remaining privately owned woodlands. In total, woodland amounts to almost 14% of land cover in Wales and the Assembly Government has the ability to increase this area either through direct intervention by adding to the public forest estate or by encouraging farmers and other landowners to plant new woodland. The Welsh Assembly Government is committed to increasing the woodland cover and an initial annual target of 1,500 hectares over the next three years is in place.

Woodlands for Wales, the Welsh Assembly Government's strategy for woodlands and trees, focuses on striking a balance between sequestration and retaining of carbon in woodlands, and substitution of more carbon-intensive substances with wood (for example in buildings or as fuel).

Annex A

Table of policies and proposals to meet the UK's carbon budgets

This Transition Plan shows how the Government will act to reduce emissions within the UK in the traded sector, where emissions are projected to fall significantly, and in the rest of the economy (the non-traded sector). Emissions in the traded sector, for the purposes of accounting under the Climate Change Act, are fixed at the level of the UK's share of the declining EU Emissions Trading System cap. This will be equal to the level of auctioning rights the UK receives plus the number of EU allowances that are freely allocated to UK installations. Combined with the emissions reductions that measures in the non-traded sector are expected to deliver shows how the UK, on central projections, will meet the first three carbon budgets.

Table A1 below sets out estimates of the emissions reductions from some of the existing measures (including many from the 2000 and 2006 Climate Change Programmes) that are already reducing emissions in the non-traded sector and which are included in the baseline emissions projections.

Tables A2 and A3 set out the emissions savings in the Non-Traded Sector from the additional measures set out in this Transition Plan. Table A4 aggregates these emissions savings and those from activities in the EU ETS, providing a summary of total projected emissions savings[3]. This is presented for the third budget in Chart A1.[1]

Chart A1
Estimated emissions savings in the third budget (2018-22)

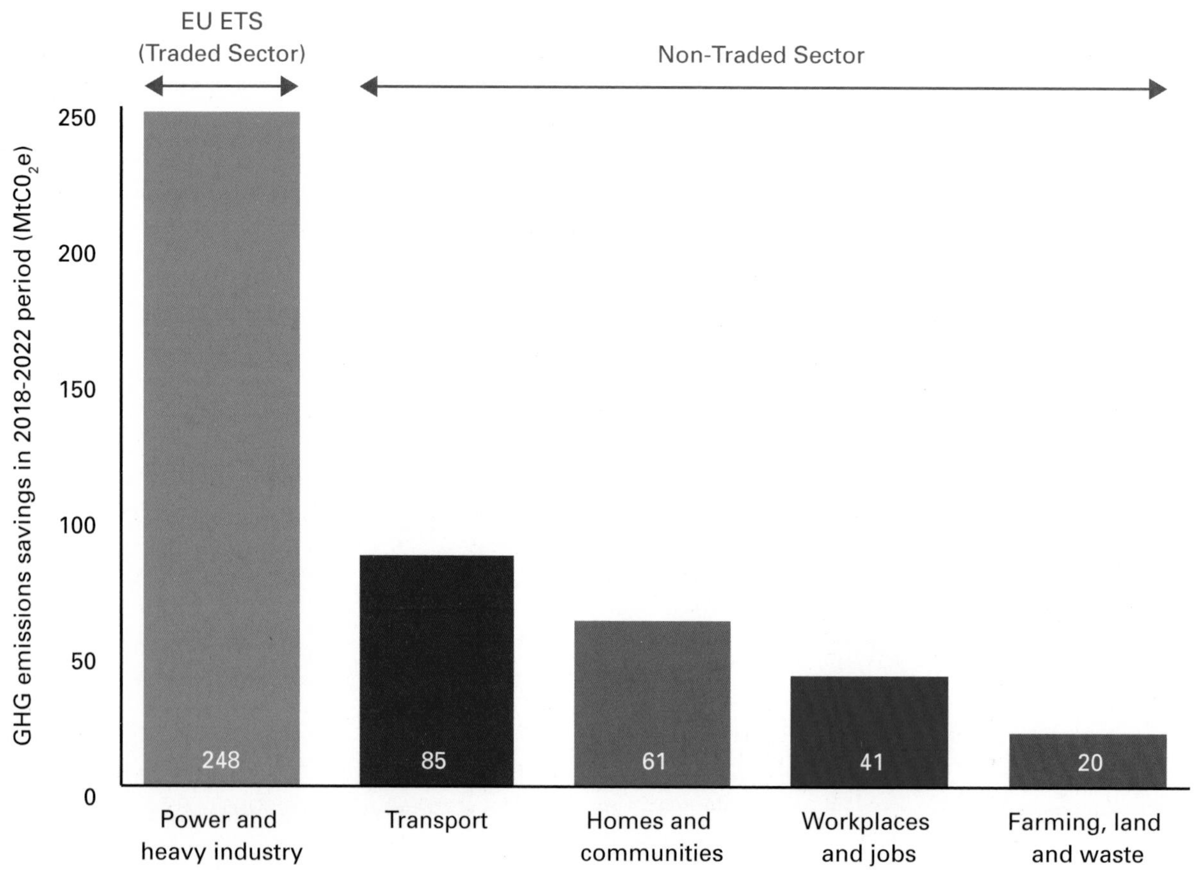

1. Emissions, for the purposes of accounting under the Climate Change Act, from activities covered by the EU Emissions Trading System (the Traded Sector) are fixed at the level of the UK's share of the EU ETS cap. This will be equal to the level of auctioning rights the UK receives plus the number of EU allowances that are freely allocated to UK installations.
Unless otherwise stated these estimates are derived from DECC's Energy Model.
The estimated carbon savings in the table are the mid-points in ranges, and represent the 'most likely' or expected outcome from delivering policy. The savings from the policies, when modelled in the DECC energy model, are slightly higher than the appraised savings from the individual policies owing to interactions within the model. The interaction effect is small relative to the volume of appraised savings – see the Analytical Annex to this Transition Plan.

Table 1

Estimated emissions savings from some of the policies and measures included in the baseline emissions projections (A)[2]

Sector	$MtCO_2e$		2008	2009	2010	2011	2012	Budget 1 (2008-12)	2013	2014	2015	2016	2017	Budget 2 (2013-17)	2018	2019	2020	2021	2022	Budget 3 (2018-22)
Homes and Communities	Energy Efficiency Commitments (2002-5 and 2005-8)[3]	*Non-traded*	1.6	1.6	1.6	1.6	1.6	**8.10**	1.6	1.6	1.6	1.5	1.5	**7.8**	1.4	1.4	1.3	1.3	1.3	**6.7**
	Building Regulations	*Non-traded*	2.7	3.2	3.7	4.5	4.9	**19.0**	5.4	5.8	6.1	6.4	6.6	**30.3**	6.9	7.1	7.2	6.9	6.5	**34.6**
	Warm Front and fuel poverty	*Non-traded*	0.8	0.7	0.7	0.7	0.7	**3.6**	0.7	0.7	0.7	0.7	0.7	**3.5**	0.7	0.7	0.7	0.7	0.7	**3.5**
Workplaces and Jobs	Building Regulations (commercial)	*Non-traded*	0.6	0.7	0.8	1.0	1.0	**4.0**	1.1	1.1	1.1	1.2	1.2	**5.6**	1.2	1.2	1.2	1.2	1.1	**5.9**
	Building Regulations (industry)	*Non-traded*	0.1	0.1	0.1	0.1	0.1	**0.5**	0.1	0.1	0.2	0.2	0.2	**0.7**	0.2	0.2	0.2	0.2	0.2	**0.8**
	Carbon Trust measures (industry)	*Non-traded*	0.0	0.0	0.0	0.1	0.1	**0.2**	0.1	0.1	0.1	0.1	0.1	**0.3**	0.1	0.1	0.1	0.1	0.1	**0.3**
	Carbon Trust measures (commercial)	*Non-traded*	0.0	1.0	0.1	0.1	0.1	**0.5**	0.1	0.1	0.1	0.1	0.1	**0.7**	0.1	0.1	0.1	0.1	0.1	**0.7**
	Climate change agreements	*Non-traded*	1.9	2.0	2.1	2.1	2.1	**10.3**	2.1	2.1	2.1	2.1	2.1	**10.6**	2.1	2.1	2.1	2.1	2.1	**10.6**
	Revolving loan fund (salix) (Public sector)	*Non-traded*	0.0	0.1	0.1	0.2	0.2	**0.5**	0.2	0.2	0.2	0.2	0.2	**0.8**	0.2	0.2	0.2	0.2	0.2	**0.8**
Transport	Renewable Transport Fuel Obligation (5% by volume)	*Non-traded*	2.5	3.0	2.9	3.8	4.5	**16.6**	5.0	5.2	5.1	5.1	5.1	**25.5**	5.1	5.0	5.0	5.0	5.0	**25.0**
	EU voluntary agreements on new car CO_2 (to 2009), including supporting fiscal measures[4]	*Non-traded*	4.3	4.7	5.1	5.6	6.1	**25.7**	6.6	7.1	7.6	7.6	7.6	**36.4**	7.6	7.7	7.7	7.7	7.7	**38.4**

2. Figures may not sum for budget periods owing to rounding.
3. Forerunner to CERT.
4. Figures may not sum for budget periods owing to rounding.

Table A2
Estimated emissions savings from the additional measures set out in this Transition Plan[7]

Sector	$MtCO_2e$		2008	2009	2010	2011	2012	Budget 1 (2008-12)	2013
Homes and Communities	Product policy[5]	*Non-traded*	-0.1	-0.1	-0.2	-0.2	-0.3	**-0.8**	-0.3
	Carbon Emission Reduction Target (2008-2011)[6]	*Non-traded*	0.8	0.9	1.7	2.5	2.6	**8.5**	2.7
	Obligation on energy suppliers[7]	*Non-traded*	0.0	0.0	0.0	0.0	0.9	**0.9**	1.7
	Community Energy Saving Programme (CESP)	*Non-traded*	-	0.0	0.0	0.0	0.0	**0.2**	0.0
	Domestic smart metering roll out	*Non-traded*	0.0	0.0	0.0	0.5	0.5	**0.9**	0.5
	Zero carbon homes[8]	*Non-traded*	0.0	0.0	0.0	0.0	0.0	**0.1**	0.1
	Renewable Heat Incentive (residential sector)	*Non-traded*	0.0	0.0	0.0	0.1	0.2	**0.3**	0.3
	TOTAL HOMES AND COMMUNITIES	*Non-traded*	0.7	0.8	1.5	2.9	3.9	**10.1**	5.0
Workplaces and Jobs	Product policy (commercial)	*Non-traded*	-0.1	-0.2	-0.2	-0.3	-0.3	**-1.1**	-0.4
	Energy Performance of Buildings Directive[9]	*Non-traded*	0.0	0.0	0.0	0.0	0.0	**0.0**	0.0
	Smart metering (small and medium business)[10]	*Non-traded*	0.0	0.0	0.0	0.0	0.1	**0.1**	0.2

5. Product policy savings are negative in the non-traded sector because of the heat replacement effect: more energy efficient products produce less ambient heat, which needs replacing via alternative fuel sources. Overall, product policy provides a significant net benefit, due to savings in emissions in the traded sector and their associated benefits.
6. The ambition for CERT was extended in the 2007 Energy White Paper, and a 20% uplift to the target was proposed in September 2008. While the savings are presented here as if they are additional to the baseline, please note that a proportion of this ambition was announced prior to the 2007 Energy White Paper.
7. This includes the savings from the proposed extension to CERT to the end of 2012.
8. Zero Carbon Homes figures presented here refer to the onsite energy efficiency elements of zero carbon homes from 2016, including the tightening of Building Regulations energy efficiency standards in 2010 and 2013. Carbon savings from on- and offsite renewable energy were removed so as not to overlap with Feed-In-Tariffs and the Renewable Heat Incentive, for which zero carbon homes would be eligible. See analytical annex for more details.
9. Includes Energy Performance Certificates, Display Energy Certificates for public buildings, inspections for air conditioning systems, and advice and guidance for boiler users.
10. Estimated savings from residential smart meters have been revised in a separate exercise, which is not reflected in the table above, please see: The impact assessment of a GB-wide smart meter roll out for the domestic sector, available from: www.decc.gov.uk/en/content/cms/consultations/smart_metering/smart_metering.aspx. The latest work suggests that in total the savings shown in the table above are broadly accurate, but more weighted to the non-traded sector and to later years than is suggested above.

2014	2015	2016	2017	Budget 2 (2013-17)	2018	2019	2020	2021	2022	Budget 3 (2018-22)
-0.4	-0.5	-0.6	-0.7	**-2.4**	-0.8	-0.9	-1.0	-1.0	-1.0	**-4.5**
2.7	2.7	2.6	2.6	**13.3**	2.4	2.2	1.9	1.8	1.7	**10.0**
2.6	3.4	4.3	5.1	**17.1**	6.0	6.8	7.7	7.7	7.7	**35.8**
0.0	0.0	0.0	0.0	**0.1**	0.0	0.0	0.0	0.0	0.0	**0.1**
0.4	0.4	0.4	0.4	**2.1**	0.4	0.4	0.3	0.3	0.3	**1.8**
0.1	0.1	0.1	0.2	**0.6**	0.3	0.3	0.4	0.5	0.6	**2.2**
0.5	0.7	1.1	1.5	**4.2**	2.0	2.8	3.6	3.6	3.6	**15.4**
5.9	6.8	7.9	9.2	**35.0**	10.3	11.6	12.9	12.9	12.9	**60.8**
-0.4	-0.5	-0.6	-0.6	**-2.5**	-0.7	-0.8	-0.8	-0.8	-0.8	**-3.9**
0.0	0.1	0.1	0.1	**0.3**	0.1	0.1	0.1	0.1	0.1	**0.7**
0.3	0.4	0.6	0.7	**2.2**	0.8	0.9	1.0	1.0	1.0	**4.7**

Sector	$MtCO_2e$		2008	2009	2010	2011	2012	Budget 1 (2008-12)	2013
Workplaces and Jobs	Carbon Reduction Commitment (commercial sector)	*Non-traded*	0.0	0.0	0.0	0.1	0.1	**0.2**	0.2
	Carbon Reduction Commitment (industry)	*Non-traded*	0.0	0.0	0.0	0.0	0.0	**0.1**	0.1
	Renewable Heat Incentive (commercial sector)	*Non-traded*	0.0	0.0	0.0	0.1	0.1	**0.2**	0.2
	Renewable Heat Incentive (industry)	*Non-traded*	0.0	0.0	0.0	0.0	0.1	**0.1**	0.1
	One-off interest free loans to SMEs	*Non-traded*	-	0.0	0.1	0.1	0.1	**0.2**	0.1
	SUB- TOTAL BUSINESS	*Non-traded*	-0.1	-0.2	-0.1	-0.0	0.2	**-0.3**	0.4
	Carbon Reduction Commitment (public sector)	*Non-traded*	0.0	0.0	0.0	0.0	0.1	**0.1**	0.1
	Renewable Heat Incentive (public sector)	*Non-traded*	0.0	0.0	0.0	0.1	0.1	**0.2**	0.2
	One-off interest free public sector loans	*Non-traded*	-	-	0.0	0.0	0.0	**0.1**	-
	SUB-TOTAL PUBLIC SECTOR	*Non-traded*	0.0	0.0	0.0	0.2	0.2	**0.3**	0.3
	TOTAL WORKPLACES	*Non-traded*	-0.1	-0.2	-0.1	0.1	0.4	**0.1**	0.7
Transport	EU new car average fuel efficiency standards of 130g CO_2/km by 2015	*Non-traded*	0.0	0.0	0.0	0.0	0.0	**0.0**	0.1
	Extension of biofuels to 10% (by energy)	*Non-traded*	0.0	0.0	0.0	0.0	0.0	**0.0**	0.0
	Low carbon emission buses	*Non-traded*	0.0	0.0	0.1	0.0	0.0	**0.0**	0.0
	SAFED training for bus drivers	*Non-traded*	0.0	0.0	0.1	0.1	0.2	**0.4**	0.2
	TOTAL TRANSPORT	*Non-traded*	0.0	0.0	0.1	0.1	0.2	**0.4**	0.3
Farming & waste	Continuation of the landfill tax escalator	*Non-traded*	0.0	0.0	0.0	0.0	0.0	**0.0**	0.2
	TOTAL	***Non-traded***	**0.6**	**0.6**	**1.5**	**3.1**	**4.5**	**10.1**	**5.8**

2014	2015	2016	2017	Budget 2 (2013-17)	2018	2019	2020	2021	2022	Budget 3 (2018-22)
0.2	0.3	0.4	0.4	**1.5**	0.5	0.5	0.6	0.6	0.6	**2.7**
0.1	0.1	0.1	0.1	**0.4**	0.1	0.2	0.2	0.2	0.2	**0.8**
0.3	0.5	0.6	0.9	**2.4**	1.3	1.6	2.3	2.3	2.3	**9.7**
0.2	0.3	0.5	0.7	**1.9**	0.9	1.2	1.5	1.6	1.6	**6.8**
0.0	0.0	0.0	0.0	**0.2**	0.0	0.0	-	-	-	**0.0**
0.7	1.2	1.6	2.3	**6.3**	3.0	3.7	4.8	4.9	4.9	**21.5**
0.1	0.1	0.1	0.2	**0.6**	0.2	0.2	0.2	0.3	0.2	**1.1**
0.3	0.6	0.6	0.9	**2.6**	1.4	1.8	2.5	2.5	2.5	**10.7**
-	-	-	-	**-**	-	-	-	-	-	**-**
0.4	0.7	0.7	1.1	**3.2**	1.6	2.0	2.7	2.7	2.7	**11.8**
1.1	1.8	2.4	3.4	**9.4**	4.6	5.7	7.6	7.6	7.6	**33.0**
0.5	1.0	1.5	2.1	**5.1**	2.7	3.4	4.0	4.7	5.3	**20.1**
0.8	1.8	2.8	3.7	**9.1**	4.7	5.6	6.6	6.6	6.6	**30.1**
0.0	0.0	0.1	0.1	**0.2**	0.1	0.1	0.2	0.2	0.3	**0.9**
0.2	0.2	0.2	0.2	**1.0**	0.2	0.2	0.2	0.2	0.2	**1.0**
1.5	3.0	4.6	6.1	**15.4**	7.7	9.3	11.0	11.7	12.4	**52.1**
0.2	0.2	0.2	0.2	**0.8**	0.3	0.3	0.3	0.3	0.3	**1.7**
8.6	**11.8**	**15.1**	**18.9**	**60.1**	**22.9**	**26.9**	**31.8**	**32.5**	**33.2**	**147.2**

Table A3

Estimated further intended emissions savings from the additional measures set out in this Transition Plan

Chapter	$MtCO_2e$		2008	2009	2010	2011	2012	**Budget 1 (2008-12)**	2013
Workplaces & Jobs	Energy intensive industries	*Non-traded*	0.0	0.0	0.0	0.0	0.0	**0.0**	1.6
Transport	Complementary measures in cars	*Non-traded*	0.0	0.0	0.0	0.1	0.2	**0.3**	0.3
	Low rolling resistance tyres for HGVs	*Non-traded*	0.0	0.0	0.0	0.0	0.0	**0.0**	0.0
	Additional impact of new car average fuel efficiency standards of 95g CO_2/km by 2020	*Non-traded*	0.0	0.0	0.0	0.0	0.0	**0.0**	0.0
	Potential EU new van CO_2 regulation	*Non-traded*	0.0	0.0	0.1	0.3	0.5	**1.0**	0.7
	Rail electrification (illustrative savings)	*Non-traded*	0.0	0.0	0.0	0.0	0.0	**0.0**	0.0
	TOTAL TRANSPORT	*Non-traded*	**0.0**	**0.0**	**0.1**	**0.4**	**0.7**	**1.2**	**1.0**
Farming, land & waste	Agriculture[11]	*Non-traded*	0.0	0.0	0.0	0.0	0.0	**0.0**[12]	0.0
	Waste[14]	*Non-traded*	0.0	0.0	0.0	0.0	0.0	**0.0**[12]	0.0
	TOTAL	***Non-traded***	**0.0**	**0.0**	**0.1**	**0.4**	**0.7**	**1.2**	**2.6**

Table A4

Total emissions savings from policies in this Transition Plan

	Budget 1 (2008-12)	**Budget 2** (2013-17)	**Budget 3** (2018-22)
TOTAL (Non-Traded Sector) (Tables A2 and A3)	12	78	208
TOTAL (EU ETS)[15] (tables A5 and A6)	0	155	248
Macroeconomic Interaction[16]	1	11	3
TOTAL PROJECTED EMISSIONS SAVINGS (Central Estimate)	**13**	**243**	**459**

2014	2015	2016	2017	**Budget 2 (2013-17)**	2018	2019	2020	2021	2022	**Budget 3 (2018-22)**
1.6	1.6	1.6	1.6	**8.0**	1.6	1.6	1.6	1.6	1.6	**8.0**
0.4	0.5	0.6	0.7	**2.6**	0.7	0.8	0.8	0.7	0.6	**3.7**
0.0	0.0	0.0	0.1	**0.1**	0.1	0.2	0.2	0.3	0.3	**1.1**
0.0	0.0	0.3	0.8	**1.0**	1.5	2.5	3.7	4.8	5.9	**18.5**
0.9	1.0	1.2	1.4	**5.2**	1.6	1.7	1.9	2.0	2.1	**9.3**
0.0	0.0	0.0	0.0	**0.0**	0.2	0.2	0.2	0.2	0.2	**0.8**
1.3	**1.5**	**2.1**	**3.0**	**8.9**	**4.1**	**5.4**	**6.8**	**8.0**	**9.1**	**33.4**
0.0	0.0	0.0	0.0	**0.0**[12]	3.0	3.0	3.0	3.0	3.0	**15.0**[13]
0.0	0.0	0.0	0.0	**0.0**[12]	0.6	0.6	0.6	0.6	0.6	**3.3**[13]
2.9	**3.1**	**3.7**	**4.6**	**16.9**	**9.3**	**10.6**	**12.0**	**13.2**	**14.3**	**59.4**

11. Emission reductions from crop management and fertiliser use, enteric fermentation and manure management.
12. There will be emissions reductions in earlier years as measures introduced to secure reductions in the third carbon budget period take hold. Because these earlier reductions aren't precisely known, the Government considers it more prudent not to include these reductions in emissions projections now.
13. Emissions reductions are intended to average to these figures across the third carbon budget period, but may not correspond to the exact figures shown in each individual year.
14. This includes reducing emissions associated with the landfilling of biogenic material such as food and wood.
15. EU Emissions Trading System Phase 3.
16. The savings from the package of policies, when modelled in the DECC energy model, are slightly higher than the appraised savings from the individual policies owing to interactions within the model. The interaction effect is small relative to the volume of appraised savings – see the Analytical Annex to this Transition Plan.

Emissions, for the purposes of accounting under the Climate Change Act, from activities covered by the EU Emissions Trading System (the Traded Sector) are fixed at the level of the UK's share of the EU ETS cap. This will be equal to the level of auctioning rights the UK receives plus the number of EU allowances that are freely allocated to UK installations.

Policies that reduce emissions in the UK in the Traded Sector will have no effect on the level of EU-wide emissions, or the net UK carbon account. They will, however, reduce the UK's net import of carbon units, with associated economic benefits. The carbon unit savings resulting from policies and measures in the baseline are set out in Table A5

The savings resulting from the proposals and policies set out in this Transition Plan are set out in Table A6.[17]

Table A5

Estimated carbon unit savings from some of the policies and measures included in the baseline emissions projections[18]

	EU Allowance Savings ($MtCO_2e$)		**2008**	**2009**	**2010**	**2011**	**2012**	**Budget 1 (2008-12)**	**2013**
Power & Heavy Industry	Renewables Obligation	*Traded*	8.8	9.9	10.4	11.1	11.9	**52.1**	12.9
Homes & Communities	Energy Efficiency Commitments (2002-5 & 2005-8)[19]	*Traded*	1.8	1.8	1.8	1.8	1.8	**8.75**	1.7
	Building Regulations	*Traded*	0.1	0.2	0.2	0.2	0.2	**0.9**	0.2
	Warm Front and fuel poverty programmes	*Traded*	1.1	1.4	1.7	1.9	1.9	**7.9**	1.9
Workplaces & Jobs	Building Regulations (commercial)	*Traded*	0.2	0.3	0.4	0.4	0.4	**1.8**	0.5
	Building Regulations (industry)	*Traded*	0.3	0.4	0.4	0.5	0.5	**2.1**	0.6
	Carbon Trust measures (industry)	*Traded*	0.1	0.1	0.3	0.3	0.3	**1.1**	0.3
	Carbon Trust measures (commercial)	*Traded*	0.1	0.1	0.2	0.2	0.2	**0.8**	0.2
	Climate Change Agreements	*Traded*	1.9	1.9	2.0	2.0	2.0	**9.7**	2.0
	Revolving loan fund (salix) (Public sector)	*Traded*	0.0	0.1	0.2	0.2	0.2	**0.8**	0.2

17. The estimated EU allowance savings given in the table are the mid-points in ranges, and represent the 'most likely' or expected outcome from delivering policy.
18. Figures may not sum for budget periods due to rounding.
19. Forerunner of CERT.

2014	2015	2016	2017	**Budget 2 (2013-17)**	2018	2019	2020	2021	2022	**Budget 3 (2018-22)**
14.1	15.2	15.9	16.6	**74.7**	17.7	18.2	18.7	19.3	19.9	**93.9**
1.6	1.5	1.3	1.1	**7.10**	0.8	0.6	0.4	0.3	0.1	**2.2**
0.2	0.2	0.2	0.2	**1.1**	0.2	0.2	0.2	0.2	0.2	**1.2**
1.9	1.9	1.9	1.9	**9.4**	1.9	1.9	1.9	1.9	1.9	**9.4**
0.5	0.5	0.5	0.5	**2.5**	0.5	0.5	0.6	0.5	0.5	**2.6**
0.6	0.6	0.6	0.6	**2.9**	0.6	0.6	0.6	0.6	0.6	**3.0**
0.3	0.3	0.3	0.3	**1.6**	0.3	0.3	0.3	0.3	0.3	**1.6**
0.2	0.2	0.2	0.2	**1.2**	0.2	0.2	0.2	0.2	0.2	**1.2**
2.0	2.0	2.0	2.0	**9.9**	2.0	2.0	2.0	2.0	2.0	**9.9**
0.2	0.2	0.2	0.2	**1.1**	0.2	0.2	0.2	0.2	0.2	**1.1**

Table A6
Estimated carbon unit savings from additional firm and funded measures[20]

Sector	$MtCO_2e$		2008	2009	2010	2011	2012	Budget 1 (2008-12)	2013
Power & Heavy Industry	Additional renewables in electricity generation from UK Renewable Energy Strategy[21]	*Traded*	0.0	0.0	0.0	0.0	0.5	**0.5**	2.6
	Carbon Capture and Storage Demonstration[22]	*Traded*	0.0	0.0	0.0	0.0	0.0	**0.0**	0.0
	TOTAL POWER AND HEAVY INDUSTRY SECTOR	***Traded***	**0.0**	**0.0**	**0.0**	**0.0**	**0.5**	**0.5**	**2.6**
Homes & Communities	Product policy	*Traded*	0.3	0.6	0.9	1.2	1.5	**4.4**	1.8
	Carbon Emission Reduction Target (2008-2011)[23]	*Traded*	0.2	1.23	1.6	1.9	1.9	**6.8**	1.9
	Obligation on energy suppliers[24]	*Traded*	0.0	0.0	0.0	0.0	0.6	**0.6**	1.1
	Community Energy Saving Programme (CESP)	*Traded*	-	0.1	0.1	0.1	0.1	**0.3**	0.1
	Domestic smart metering roll out	*Traded*	0.0	0.0	0.0	1.3	1.3	**2.5**	1.3
	Zero carbon homes[25]	*Traded*	0.0	0.0	0.0	0.0	0.1	**0.1**	0.1
	TOTAL HOMES AND COMMUNITIES	***Traded***	**0.5**	**2.0**	**2.6**	**4.4**	**5.4**	**14.7**	**6.4**

20. Figures may not sum for budget periods due to rounding.
21. Includes impact of the increase in, and extension of, the Renewables Obligation, introduction of the feed-in tariff, and other supporting measures.
22. The carbon savings presented here are consistent with the updated emissions projections but do not fully reflect the proposals in the June consultation: A framework for the development of clean coal Available from: www.decc.gov.uk/en/content/cms/consultations/clean_coal/clean_coal.aspx.
23. The ambition for CERT was extended in the 2007 Energy White Paper, and a 20% uplift to the target was proposed in September 2008. While the savings are presented here as if they are additional to the baseline, please note that a proportion of this ambition was announced prior to the 2007 Energy White Paper.
24. This includes the savings from the proposed extension to CERT to the end of 2012.
25. Zero Carbon Homes figures presented here refer to the onsite energy efficiency elements of zero carbon homes from 2016, including the tightening of Building Regulations energy efficiency standards in 2010 and 2013. Carbon savings from on- and offsite renewable energy were removed so as not to overlap with Feed-In-Tariffs and the Renewable Heat Incentive, for which zero carbon homes would be eligible. See analytical annex for more details.

2014	2015	2016	2017	Budget 2 (2013-17)	2018	2019	2020	2021	2022	Budget 3 (2018-22)
5.4	8.9	12.8	16.2	**45.8**	19.6	23.2	27.5	28.7	28.4	**127.4**
0.1	1.0	2.2	2.2	**5.4**	2.8	4.3	4.6	4.6	4.6	**20.9**
5.5	**9.9**	**14.9**	**18.3**	**51.2**	**22.4**	**27.6**	**32.1**	**33.3**	**33.0**	**148.3**
2.2	2.5	2.8	3.2	**12.5**	3.6	3.9	4.3	4.3	4.3	**20.2**
1.9	1.9	2.0	2.1	**9.8**	2.1	2.3	2.5	2.6	2.6	**12.00**
1.7	2.2	2.8	3.3	**11.1**	3.9	4.4	5.0	5.0	5.0	**23.3**
0.1	0.1	0.1	0.1	**0.3**	0.1	0.1	0.1	0.1	0.1	**0.3**
1.3	1.3	1.3	1.3	**6.4**	1.2	1.2	1.2	1.2	1.3	**6.1**
0.1	0.1	0.2	0.2	**0.7**	0.2	0.3	0.3	0.3	0.3	**1.4**
7.2	**8.3**	**9.2**	**10.0**	**40.8**	**11.5**	**12.2**	**13.4**	**13.4**	**13.5**	**63.3**

Sector	$MtCO_2e$		2008	2009	2010	2011	2012	Budget 1 (2008-12)	2013
Workplaces & Jobs	Product policy (commercial)	*Traded*	0.3	0.5	0.8	1.0	1.2	**3.8**	1.4
	Energy Performance of Buildings Directive[26]	*Traded*	0.0	0.0	0.0	0.0	0.1	**0.1**	0.1
	Smart metering (small and medium business)[27]	*Traded*	0.0	0.0	0.0	0.0	0.0	**0.0**	0.0
	Carbon Reduction Commitment (commercial sector)	*Traded*	0.0	0.0	0.0	0.1	0.2	**0.3**	0.3
	Carbon Reduction Commitment (industry)	*Traded*	0.0	0.0	0.0	0.1	0.2	**0.4**	0.4
	Renewable Heat Incentive (commercial sector)	*Traded*	0.0	0.0	0.0	0.0	0.0	**0.0**	0.0
	Renewable Heat Incentive (industry)	*Traded*	0.0	0.0	0.0	0.1	0.2	**0.3**	0.4
	One-off interest free loans to SMEs[28]	*Traded*	-	0.0	0.1	0.1	0.1	**0.2**	0.1
	SUB- TOTAL BUSINESS	*Traded*	0.3	0.5	0.9	1.5	2.1	**5.1**	2.6
	Carbon Reduction Commitment (public sector)	*Traded*	0.0	0.0	0.0	0.0	0.1	**0.1**	0.1
	Product policy (public sector)[29]	*Traded*	0.0	0.1	0.1	0.1	0.1	**0.4**	0.2
	One-off interest free public sector loans[30]	*Traded*	-	-	0.0	0.0	0.0	**0.1**	-
	SUB-TOTAL PUBLIC SECTOR	***Traded***	**0.0**	**0.1**	**0.1**	**0.2**	**0.2**	**0.6**	**0.3**
	TOTAL WORKPLACES	***Traded***	**0.3**	**0.6**	**1.0**	**1.6**	**2.3**	**5.7**	**2.9**
	TOTAL	***Traded***	**0.8**	**2.6**	**3.6**	**6.1**	**8.3**	**21.3**	**11.9**

2014	2015	2016	2017	Budget 2 (2013-17)	2018	2019	2020	2021	2022	Budget 3 (2018-22)
1.6	1.8	2.0	2.3	**9.1**	2.5	2.8	3.0	3.0	3.0	**14.2**
0.1	0.1	0.2	0.2	**0.7**	0.3	0.3	0.3	0.3	0.3	**1.5**
0.1	0.1	0.1	0.1	**0.4**	0.2	0.2	0.2	0.2	0.2	**0.9**
0.5	0.6	0.7	0.8	**2.9**	0.9	1.0	1.2	1.2	1. 2	**5.4**
0.5	0.6	0.7	0.8	**2.9**	0.9	1.0	1.2	1.2	1.2	**5.4**
0.0	0.0	0.1	0.1	**0.2**	0.1	0.2	0.2	0.2	0.2	**1.0**
0.6	0.9	1.6	2.2	**5.8**	2.8	3.8	4.7	4.7	4.7	**20.7**
0.0	0.0	0.0	0.0	**0.2**	0.0	0.0	-	-	-	**0.0**
3.4	4.1	5.4	6.5	**22.1**	7.7	9.3	10.7	10.7	10.7	**49.1**
0.2	0.2	0.2	0.3	**1.0**	0.3	0.4	0.4	0.4	0.4	**1.8**
0.2	0.2	0.2	0.3	**1.0**	0.3	0.3	0.3	0.3	0.3	**1.6**
-	-	-	-	**-**	-	-	-	-	-	**-**
0.3	**0.4**	**0.5**	**0.5**	**2.0**	**0.6**	**0.7**	**0.7**	**0.7**	**0.7**	**3.4**
3.7	**4.5**	**5.8**	**7.0**	**24.1**	**8.3**	**9.9**	**11.5**	**11.4**	**11.4**	**52.5**
16.6	**22.8**	**30.3**	**35.9**	**117.4**	**42.2**	**50.2**	**57.4**	**58.8**	**58.6**	**267.2**

26. Includes Energy Performance Certificates, Display Energy Certificates for public buildings, inspections for air conditioning systems, and advice and guidance for boiler users.
27. Estimated savings from residential smart meters have been revised in a separate exercise, which is not reflected in the table above, please see: The impact assessment of a GB-wide smart meter roll out for the domestic sector, available from: www.decc.gov.uk/en/content/cms/consultations/smart_metering/smart_metering.aspx. The latest work suggests that in total the savings shown in the table above are broadly accurate, but more weighted to the non-traded sector and to later years than is suggested above.
28. Announced at Budget 2009.
29. The effect of product policy on the non-traded sector is not shown as the heat replacement effect for the public sector is covered by the CRC cap. See also footnote 7.
30. Announced at Budget 2009.

Indicative Annual Ranges

Having set the carbon budget for a budgetary period, section 12 of the Climate Change Act requires the Government to publish indicative annual ranges showing where it expects 'the net UK carbon account' to fall in each year of the budgetary period. This annex meets the requirement of section 12 in respect of the first three budgetary periods.

The net UK carbon account is what the Government compares against the carbon budgets to determine whether they are being met – it must not exceed the carbon budget at the end of each budgetary period. It is calculated by first taking net UK emissions (i.e. aggregate gross emissions from sources in the UK, adjusted to take into account removals by sinks). These are adjusted to account for any carbon units which have been brought in from overseas by Government and others to offset UK emissions ('credits'), and UK carbon units which have been disposed of to a third party ('debits').

As the Government plans not to use credits in the non-traded sector to meet carbon budgets, the expectation for the net UK carbon account corresponds to the projected net UK emissions, taking into account the expected level of credits or debits due to the use or disposal of carbon units by UK participants in the EU Emissions Trading System.

The indicative annual ranges for the first three carbon budgets are shown in Chart A2 and Table A7 below. The levels have been set with reference to the uncertainty ranges based on the central economic growth forecast in the latest emissions projections (discussed in the separate analytical annex and emissions projections publication).[31] It can therefore be expected, with a high degree of confidence (95%), that the net UK carbon account will fall within the range for each year.

The Government considers it very unlikely that the account will be at the upper bound of the range in each year, and these ranges should not be interpreted as an indication that this would be satisfactory. Falling at, or near, the upper bound in each year would be incompatible with meeting the carbon budgets, even taking into account the provision to borrow from the next budget period, and may also be incompatible with compliance with EU obligations.[32] Further action may be required if the net UK carbon account were at or near the upper bound in any year to bring about the necessary reductions in future levels of the account. The Government's general approach to dealing with this uncertainty is set out in chapter 2 and discussed in more detail in the analytical annex to this Plan.

The annual statement of emissions required under section 16 of the Climate Change Act will set out the actual net UK carbon account for each year. This will show whether the account is within the ranges identified in this annex. The first such statement, in respect of the net UK carbon account for 2008, must be published by 31 March 2010.

31. For further details please see www.decc.gov.uk/en/content/cms/publications/lc_trans_plan/lc_trans_plan.aspx
32. The exact details of the UK's obligations under the EU package, in both the traded (EU Emissions Trading System) and non-traded (outside the EU ETS) sectors cannot yet be determined as they are dependent on future emissions.

Projected sectoral contribution to the net UK carbon account

Chart A3 plots projected UK greenhouse gas emissions by sources minus removals by sinks against the projected net UK carbon account, to show the expected contribution of emissions trading under the EU Emissions Trading System to meeting carbon budgets. This illustrates that the Government expects UK participants in the EU ETS to vary over the three budget periods between being net sellers and net purchasers of carbon units from abroad."

Table A8 shows the expected contribution to the net UK carbon account by sector (using the sectorial split used in this Transition Plan). Table A9 shows the same information, but splits the sectors in the way in which the Government reports in the National Communication to the UNFCCC.[33]

Chart A2

Indicative annual ranges for the net UK carbon account

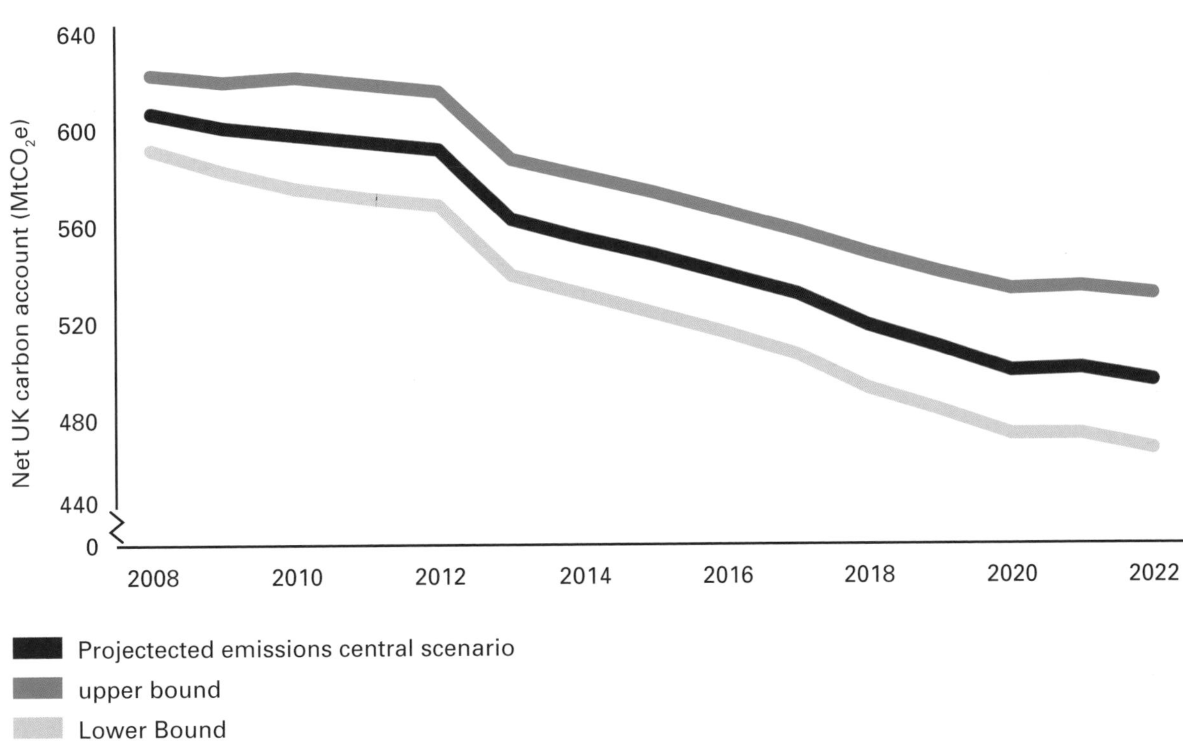

33. Available from: www.decc.gov.uk/en/content/cms/what_we_do/change_energy/the_issue/strategy/strategy.aspx

Table A7

Indicative annual ranges for the net UK carbon account

Net UK carbon account ($MtCO_2e$)	2008	2009	2010	2011	2012	2013	2014	2015	2016	2017	2018	2019	2020	2021	2022
Upper bound	619	616	618	615	612	584	577	570	562	554	545	537	530	531	528
Projected emissions (central scenario)	603	597	594	591	588	559	551	544	536	528	515	506	496	497	492
Lower bound	588	579	572	568	565	536	528	520	512	503	489	480	470	470	464

Table A8

Expected Contribution to the Net UK Carbon Account from Sectors as defined in this Transition Plan

$MtCO_2e$	2008	2009	2010	2011	2012	2013	2014	2015	2016	2017	2018	2019	2020	2021	2022
Power and heavy industry	246	246	246	246	248	224	220	216	212	208	203	199	195	195	191
Homes and communities	84	83	81	78	76	73	71	69	67	64	63	61	60	60	61
Workplaces and jobs	74	72	70	69	66	64	64	64	64	64	62	61	59	60	60
Transport[34]	130	126	126	126	126	125	124	122	121	119	117	115	112	111	110
Farms, land and waste	70	70	71	71	72	72	72	73	73	73	69	69	70	70	70
TOTAL (net UK carbon account)	603	597	594	591	588	559	551	544	536	528	515	506	496	497	492

34. These projections may differ marginally from the estimates in the Carbon Reduction Strategy for Transport, due to both the inclusion of military aviation and shipping emissions in the projections provided here, and the use of the DECC Energy Model (rather than the DfT's National Transport Model) to derive the results. See http://www.dft.gov.uk/pgr/economics/ntm/roadtransportforcasts08/rtf08.pdf for more detail about the differences between the two models.

Table A9

Expected contribution to the net UK carbon account from sectors on the basis of the UK's reporting to the UNFCCC

MtCO2e	2008	2009	2010	2011	2012	2013	2014	2015	2016	2017	2018	2019	2020	2021	2022
Energy supply	217	203	190	191	186	184	173	172	164	158	155	145	134	128	125
Business	93	90	87	86	86	85	87	89	90	91	92	92	92	94	94
Industrial processes	19	17	16	16	17	17	17	17	17	17	17	17	17	16	16
Transport	130	126	126	126	126	125	124	122	121	119	117	115	112	111	110
Residential	84	83	80	77	75	73	71	68	66	64	62	61	59	60	60
Public	11	11	10	10	10	10	10	10	10	10	9	9	8	8	8
Agriculture	49	49	50	50	50	50	50	50	50	50	47	47	47	47	47
LULUCF (net)	-2	-2	-1	-1	0	0	0	1	1	2	2	2	3	3	3
Waste management	23	23	23	22	22	22	22	22	21	21	20	20	20	20	20
TOTAL (greenhouse gas emissions by sources minus removals by sinks)	**624**	**599**	**581**	**579**	**572**	**565**	**554**	**552**	**540**	**531**	**520**	**507**	**491**	**486**	**483**
Emissions reductions or increases resulting from purchases or sales through EU ETS[35]	20	2	-13	-12	-16	6	2	7	4	3	6	1	-5	-11	-8
TOTAL (net UK carbon account)	**603**	**597**	**594**	**591**	**588**	**559**	**551**	**544**	**536**	**528**	**515**	**506**	**496**	**497**	**492**

35. A positive figure means a net amount of carbon units have been purchased by UK operators in the EU ETS, while a negative figure denotes a net amount of carbon units have been sold by UK operators in the EU ETS.

Chart A3

Comparison of projections of net UK carbon account and net UK greenhouse gas emissions

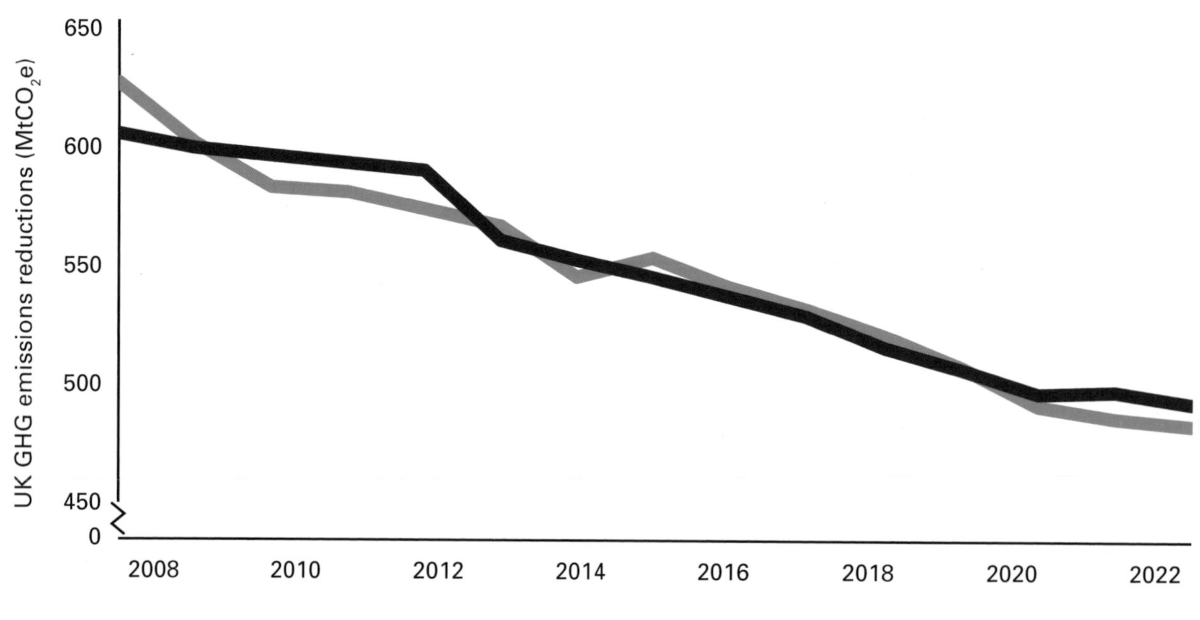

Net UK carbon account

Net UK greenhouse gas emissions
(greenhouse gas emissions from sources in the UK minus removals by sinks)

Annex B
Departmental carbon budgets

As set out in chapter 2, to ensure that every part of government takes responsibility for delivery of the UK's carbon budgets, the Government is introducing a system of departmental carbon budgets. This new approach will be the first time any government has introduced such an accountability mechanism, and will serve as a pilot that will be reviewed ahead of the second budget period (2013-2017).

Under this new system, each government Department will hold its own carbon budget made up of two elements:

a. one representing its relative **degree of influence on reducing emissions from each sector of the economy; and**

b. one reflecting the **emissions from the part of the public sector that it has responsibility for**.

To determine each Department's relative degree of influence over the various parts of the economy, a broad set of factors has been taken into consideration:

a. Where a Department holds policy levers for reducing greenhouse gas emissions, for which a specific projected impact on cutting emissions can be identified (for instance, regulations on building or product standards);

b. Responsibility for policies, policy areas and economic activity which result in increases in emissions, whether as a direct result of economic activity or of creating demand for that activity (for example, increased transport emissions as a result of the school run or freight haulage); and

c. The overarching responsibility and influence of particular government Departments with respect to those areas of the economy which they most closely sponsor, and for whom they are the main point of contact in Government.

This inevitably results in a broad approximation, rather than an exact measurement, of a Department's relative degree of influence over a sector's emissions. The budgets are not designed to demonstrate the precise contribution that a Department will make to reducing emissions from a particular sector.

Instead, analogous with Public Service Agreements, their purpose is to give Departments a stake in reducing emissions from a given sector. Importantly, the only way that a Department can deliver its carbon budget is to work with other Departments, to ensure that emissions from each sector in which it has a stake are reduced to the levels required to meet the UK's carbon budget.

With respect to emissions from the wider public sector, the intention is that over time the Government will seek to include in Departments' carbon budgets the emissions from the parts of the public sector that fall within their respective areas of responsibility.

As a first step the departmental carbon budgets shown below include emissions from central government, based on the existing Sustainable Operations on the Government Estate framework and a target to reduce CO2 emissions from office buildings by 30% by 2020 from a 1999-2000 baseline; and from Departments' own administrative transport, by 30% from a 2005-06 baseline.

The remainder of emissions from the public sector have, for the moment, been included in the Department of Energy and Climate Change carbon budget, given its policy responsibility for the Carbon Reduction Commitment. However, the Government's intention is to include emissions from schools, the NHS and further and higher education institutions in the relevant Departments' carbon budgets by April 2010.

The table overleaf sets out the carbon budgets for each major Whitehall Department in each of the first three carbon budget periods.

At the end of the first budget period, each Department's outturn against its carbon budget will be assessed based on its share of responsibility for emissions from each sector of the economy, and how well that sector has succeeded in reducing greenhouse gas emissions.

If, in future, unexpected changes in circumstances mean that there is a shortfall against the UK's carbon budgets, the Government may need to purchase international credits in order to meet its obligations under the Climate Change Act, effectively imposing a financial liability on the Government. This means that there could be real financial consequences for Government if it does not achieve its carbon budgets through domestic emissions reductions alone.

The departmental carbon budgets will be the subject of review over the first budget period, looking at a number of factors:

a. How well these budgets support the effective operation of a system to manage the UK's carbon budgets;

b. How well the initial allocations reflect the real influence that each Department has for each sector of the economy;

c. How best to take account of the extent to which responsibility for different sectors of the economy is devolved in Scotland, Wales and Northern Ireland, which will require further analysis and discussion with the devolved administrations; and

d. How best to capture the emission reductions that we want to deliver from across every part of the public sector.

	Homes and communities 2008–2012		Transport 2008–2012		Waste 2008–2012		Power and heavy industry 2008–2012		Workplaces and jobs 2008–2012 Industrial process	
	%	$MtCO_2e$	%	$MtCO_2e$	%	$MtCO_2e$	%	$MtCO_2e$	%	$MtCO_2e$
Department of Energy and Climate Change	63	257	1	6	7	8	100	1011	51	43
Department for Children Schools and Families	0	0	1	6	0	0	0	0	0	0
Ministry of Defence	0	0	3	19	0	0	0	0	0	0
Department for Health	0	0	1	6	0	0	0	0	0	0
Business, Innovation and Skills	9	37	9	58	15	17	0	0	19	16
Department of Environment Food and Rural Affairs	1	4	1	6	70	80	0	0	24	20
Department of Work and Pensions	0	0	0	0	0	0	0	0	0	0
HM Treasury[45]	0	0	0	0	0	0	0	0	0	0
Communities and Local Government	27	110	4	26	7	8	0	0	0	0
Department for Transport	0	0	76	493	1	1	0	0	6	5
Ministry of Justice	0	0	0	0	0	0	0	0	0	0
Department of Culture Media and Sport	0	0	4	26	0	0	0	0	0	0
Cabinet Office	0	0	0	0	0	0	0	0	0	0
Department for International Development	0	0	0	0	0	0	0	0	0	0
Foreign and Commonwealth Office	0	0	0	0	0	0	0	0	0	0
HM Revenue and Customs	0	0	0	0	0	0	0	0	0	0
Home Office	0	0	0	0	0	0	0	0	0	0
Law Officers	0	0	0	0	0	0	0	0	0	0
TOTAL	**100**	**408**	**100**	**648**	**100**	**115**	**100**	**1011**	**100**	**85**

Heating workplaces		Farming and land 2008–2012		Public sector 2008–2012 (Mt CO_2e)	Public sector 2013–2017 (Mt CO_2e)	Public sector 2018–2022 (Mt CO_2e)	allocation 2008–2012 ($MtCO_2e$)	allocation 2013–2017 ($MtCO_2e$)	allocation 2018–2022 ($MtCO_2e$)	%age share of total carbon budget period 3 (%)
%	$MtCO_2e$	%	$MtCO_2e$							
80	361	2	5	39.47	37.48	30.2	1731.38	1542	1358.2	53
0	0	2	5	0.07	0.07	0.06	11 .49	11.44	11.11	0
0	0	0	0	9.51	8.95	7.61	28.96	27.18	24.8	1
1	5	2	5	0.05	0.04	0.04	15.99	15.88	15.76	1
15	68	2	5	0.07	0.07	0.06	201 .18	191.74	186.53	7
2	9	88	217	0.19	0.17	0.15	337.8	343.76	347.17	14
0	0	0	0	0.99	0.89	0.79	0.99	0.89	0.79	0
0	0	0	0	0.04	0.03	0.03	0.04	0.03	0.03	0
1	5	2	5	0.14	0.12	0.11	153.83	135.74	123.25	5
0	0	2	5	0.16	0.14	0.13	503.92	482.41	447.17	18
0	0	0	0	0.76	0.68	0.6	0.76	0.68	0.60	0
1	5	0	0	0.02	0.02	0.02	30.47	29.27	27.62	1
0	0	0	0	0.03	0.03	0.03	0.03	0.03	0.03	0
0	0	0	0	0.01	0.01	0.01	0.01	0.01	0.01	0
0	0	0	0	0.06	0.05	0.05	0.06	0.05	0.05	0
0	0	0	0	0.88	0.8	0.71	0.88	0.80	0.71	0
0	0	0	0	0.15	0.13	0.12	0.15	0.13	0.12	0
0	0	0	0	0.07	0.06	0.05	0.07	0.06	0.05	0
100	**452**	**100**	**247**	**53**	**49**	**41**	**3018**	**2782**	**2544**	**100**

Top tips for reducing your carbon footprint

Household savings:

Won't cost you a penny – do this week!

1. Switch off appliances when not in use to save £30 per year.
2. Only boiling as much water as you need could save you up to £25 a year.
3. In centrally-heated houses, try turning your thermostat down by 1ºC. Provided you are still comfortable, you could reduce CO_2 emissions and cut your fuel bill by up to 10%.

Will cost a little but you will save money almost immediately – do within the next few weeks

4. Swap traditional light bulbs with energy saving ones and you will save £60 over the lifetime of the bulb.
5. Fitting your hot water tank with an insulating jacket will only cost a few pounds and, with all the heat it traps, it pays for itself within six months. Fit one that's at least 75mm (3 inches) thick and you could save around £30 a year.

Will involve upfront costs, but you will save money over the longer term

6. Fitting loft insulation to the recommended amount (270mm) could save you up to £100 a year. Even if you already have insulation you could still save up to £30 per year by topping up.
7. Cavity wall insulation can take a matter of hours to install and could save you £150 a year on fuel bills.
8. Installing draught excluders where there are gaps could save you £20 a year.
9. If your house does not yet have double glazing, installing it could save up to £80 a year.
10. Replace a boiler (10-15 years old) with an energy efficient condensing boiler ('A' rated) and suitable controls (e.g. thermostats) could save you up to £90 each year.
11. Replace white goods with energy saving recommended appliances, and you could save between £5 and £20 each year.

Printed in the UK for The Stationery Office Limited on behalf of the Controller of Her Majesty's Stationery Office

ID 6176880 07/09

Printed on Paper containing 75% recycled fibre content minimum.

Designed by SPY Design and Publishing Ltd.
www.spydesign.co.uk